SpringerBriefs in Research and Innovation Governance

SpringerBriefs in Research and Innovation Governance present concise summaries of cutting-edge research and practical applications across a wide spectrum of governance activities that are shaped and informed by, and in turn impact research and innovation, with fast turnaround time to publication. Featuring compact volumes of 50 to 125 pages, the series covers a range of content from professional to academic. Monographs of new material are considered for the SpringerBriefs in Research and Innovation Governance series. Typical topics might include: a timely report of state-of-the-art analytical techniques, a bridge between new research results, as published in journal articles and a contextual literature review, a snapshot of a hot or emerging topic, an in-depth case study or technical example, a presentation of core concepts that students and practitioners must understand in order to make independent contributions, best practices or protocols to be followed, a series of short case studies/ debates highlighting a specific angle. SpringerBriefs in Research and Innovation Governance allow authors to present their ideas and readers to absorb them with minimal time investment. Both solicited and unsolicited manuscripts are considered for publication.

Doris Schroeder · Kate Chatfield ·
Saleh Aljadeeah · Hazel Partington ·
Michelle Singh · Orla Drummond · Young Su Park ·
Lisa Tambornino

Research Ethics and Integrity in Fragile Settings

Challenges, Complexity and a Global Code of Conduct

Springer

Authors
See next page

ISSN 2452-0519 ISSN 2452-0527 (electronic)
SpringerBriefs in Research and Innovation Governance
ISBN 978-3-032-36140-0 ISBN 978-3-032-36141-7 (eBook)
https://doi.org/10.1007/978-3-032-36141-7

This work was supported by University of Central Lancashire.

This Springer imprint is published by the registered company Springer Nature Switzerland AG
The registered company address is: Gewerbestrasse 11, 6330 Cham, Switzerland

Doris Schroeder
Centre for Professional Ethics
University of Lancashire
Preston, UK

Saleh Aljadeeah
Institute of Tropical Medicine in Antwerp
Antwerp, Belgium

Michelle Singh
European and Developing Countries
Clinical Trials Partnership Association
Cape Town, South Africa

Young Su Park
History of Medicine and Medical
Humanities
Seoul National University
Seoul, Korea (Republic of)

Kate Chatfield
Centre for Professional Ethics
University of Lancashire
Preston, UK

Hazel Partington
Centre for Professional Ethics
University of Lancashire
Preston, UK

Orla Drummond
Trilateral Research IE
Waterford, Ireland

Lisa Tambornino
EUREC Office gUG, Office
of the European Network of Research
Ethics Committees
Bonn, Germany

Foreword

Across the world, researchers are increasingly called upon to work in environments marked by fragility: regions affected by natural disasters, poverty and armed conflicts. These contexts are the ones where solid research is most urgently needed. Solid evidence can help guide humanitarian action, support recovery and reconstruction, strengthen democratic governance, and contribute to long-term resilience and stability. Yet precisely because these environments are characterised by instability, vulnerability and unequal power relations, they also raise some of the most complex ethical and integrity challenges in contemporary research.

Conducting research in fragile settings requires more than methodological rigour. It needs deep ethical reflection, strong legal and ethical governance safeguards, and a real commitment to protecting the dignity, rights and safety of the people and communities involved. Researchers may face weak institutional settings, frail oversight mechanisms and increased security risks. Obtaining meaningful informed consent, protecting sensitive personal data, ensuring the safety of research teams and participants, and avoiding unintended harm can become particularly difficult under such circumstances. These realities demand ethical guidance that is sensitive to context, attentive to power dynamics and responsive to the lived realities of those most affected.

The European Union has long recognised that excellence in research must go hand in hand with responsibility and integrity. As one of the largest public funders of research and innovation worldwide, the European Commission has consistently promoted high ethical standards as a core component of its research policy. Through successive research framework programmes, the EU has supported responsible and ethically sound research, not only in laboratories and universities but also in communities around the globe, including regions facing profound social, political and environmental challenges.

The EU has also taken important initiatives to strengthen ethical governance in international research collaborations. One landmark initiative has been the support of The TRUST Code—A Global Code of Conduct for Equitable Research Partnerships (TRUST 2018) to address a practice known as "ethics dumping", which refers to the carrying out, in settings where ethical oversight may be a challenge, of research practices that would not be considered acceptable in well-regulated environments. By articulating clear values—fairness, respect, care and honesty—the TRUST Code has provided a practical and influential framework for researchers, institutions and funders seeking to build equitable research partnerships across global power imbalances.

The impact of the TRUST Code has been considerable, but as the landscape of research continues to evolve, new and more complex challenges emerge. While the TRUST Code has been highly successful in addressing ethical issues in international partnerships, it does not fully address the complexities associated with research conducted in fragile settings. Post-disaster areas, conflict or post-conflict zones and high-crime informal settlements present circumstances that differ fundamentally from those assumed in many traditional research ethics frameworks. In such environments, institutional protections may be limited, communities may face acute vulnerabilities, and the risks associated with research activities can be heightened.

The TRUST Code Supplement presented in this book addresses specifically this gap. By extending the principles of the TRUST Code to research conducted in fragile settings, it provides additional guidance tailored to some of the most challenging research environments in the world. In doing so, it strengthens the ethical foundations upon which responsible research must be grounded.

For research funders such as the European Commission, this development is particularly important. Funding organisations have a special responsibility not only to support excellent and impactful research but also to ensure that such research is conducted responsibly wherever it takes place. As EU-funded research increasingly engages with fragile contexts—whether in humanitarian crises, disaster recovery, urban vulnerability or conflict-affected regions—the need for robust ethical frameworks becomes even more evident. The TRUST Code Supplement therefore provides an important tool to support responsible decision-making by researchers, ethics committees and funders alike.

What makes this book particularly valuable is not only the guidance it offers but also that it shows the rigorous process through which that guidance was developed. The chapters document the evidence and reflections underpinning the supplement: case studies drawn from real-world research experiences in fragile contexts, systematic scoping reviews of the ethical and integrity challenges identified in the academic literature, and insights gathered from researchers with direct experience working in such settings. This empirical grounding strengthens the credibility and practical relevance of the guidance.

At the same time, the book situates these findings within the broader conversation about responsible research and innovation. The ethical challenges of fragile settings cannot be separated from the rapidly evolving global research landscape. Emerging technologies such as artificial intelligence, advanced data analytics and digital platforms are transforming research methods and creating unprecedented opportunities to generate knowledge. However, these technologies can also introduce new risks—particularly in fragile environments where regulatory and ethics review frameworks may be weaker and vulnerabilities more pronounced. Questions of data protection, algorithmic bias, surveillance and misuse of sensitive information assume greater urgency.

For that reason, the importance of this book extends far beyond a single academic field. It is not only an important contribution to the scholarly literature on responsible research and innovation and research ethics. It is also a vital resource for the wider research ecosystem and the ethical governance structures across the world. Researchers and ethics committees working in fragile settings will find practical guidance for navigating difficult ethical dilemmas. Ethics committees and institutional review boards will gain insights into challenges that traditional ethics review processes may not fully capture. Research funders will benefit from a clearer understanding of the ethical considerations that must inform funding decisions and oversight mechanisms. Policymakers, meanwhile, will find valuable input for shaping the regulatory frameworks and governance models that are increasingly needed to guide responsible research in a rapidly changing technological and geopolitical environment.

In this sense, the book occupies an important place not only in academic discourse but also in the growing body of policy playbooks that support global governance in research and innovation. As research becomes more global, more interdisciplinary and more technologically driven, the need for shared ethical frameworks becomes ever more pressing. The TRUST Code and its Supplement represent important milestones in an international environment that is increasingly at risk of becoming more fragile and fragmented than ever before.

It is therefore my hope that the TRUST Code Supplement will follow the successful path of its big sister, the TRUST Code, and become widely used by researchers, institutions and policymakers across the world. By providing guidance that is both principled and practical, it has the potential to shape how research needs to be conducted in some of the most complex and sensitive environments.

Finally, special thanks are due to the authors of this important work. Developing ethical guidance that is both rigorous and practical requires not only scholarly expertise but also dedication, collaboration and a deep sense of responsibility toward the communities affected by research. I would like to express my appreciation to Prof. Doris Schroeder for her leadership, her unwavering commitment, and her passion for making the world of research a better and more responsible one. Her efforts, together with those of the contributors to this volume and all those who have been part of the consultations, have resulted in a work that will serve researchers, institutions and policymakers alike for years to come.

At a time when global challenges are becoming more complex and interconnected and the technological environment more disruptive than ever, the need for responsible research becomes more important than before. This book makes a significant contribution toward ensuring that research conducted in fragile settings upholds the highest standards of ethics and integrity and that it ultimately serves the people and communities whose lives it seeks to improve.

Dr. Mihalis Kritikos

References

Kritikos M (2018) EU policy making on GMOs: the false promise of proceduralism. Palgrave Macmillan, London

Kritikos M, Iphofen R (2023) Ethical AI surveillance in the workplace, Advances in Research Ethics and Integrity 10. Emerald Publishing, Leeds. https://doi.org/10.1108/S2398-6018202310

TRUST (2018) The TRUST Code: a global code of conduct for equitable research partnerships. https://doi.org/10.48508/GCC/2018.05

Dr. Mihalis Kritikos is the Secretary of the European Group on Ethics in Science and New Technologies and Senior Policy Analyst at the Ethics and Integrity Sector of the European Commission (DG-Research and Innovation) working on the ethical development of emerging technologies with special emphasis on AI ethics. He is also the author of two books in the field of disruptive technologies (Kritikos 2018; Kritikos and Iphofen 2023).

Acknowledgements This book tells the story of the TRUST Code Supplement, a story that began with the TRUST Consortium. Hence, our first thanks go to the TRUST team members who developed the TRUST Code—A Global Code of Conduct for Equitable Research Partnerships, which was launched at the European Parliament in 2018. This code has since achieved significant recognition and uptake around the world.

As use of the TRUST Code expanded, an important gap was identified, namely, the need for specialised ethics and integrity guidance tailored to fragile settings, including conflict zones, natural disaster zones and other highly unstable environments. Although research in such contexts should aspire to achieve the ethics and equity standards promoted by the TRUST Code, these settings present unique challenges. Addressing these challenges, in turn, requires specialist guidance.

We are deeply grateful to the European Commission (EC), UK Research and Innovation (UKRI) and the Swiss State Secretariat for Education, Research and Innovation (SERI) for funding the research and global consultations that made the TRUST Code Supplement possible (grant agreement 101058094).[1]

Our sincere thanks go to Dr. Mihalis Kritikos, our Policy Officer at the EC for his steadfast support and for contributing an inspiring Foreword. We are equally grateful to our Project Officer, Roberta Monachello; we were fortunate to work with her on both the TRUST Code and the TRUST Code Supplement. Important strategic input from the EC was also received from Dorian Karatzas, the former Head of Ethics and Integrity at the EC, and Dr. Filipa Ferraz de Oliveira, the Ethics Principal Agent of the European Research Council Executive Agency. At SERI, we thank Raphael Misteli and Brita Bamert very much.

We are also indebted to the PREPARED team. Although their primary task was to develop an ethics and integrity code for future pandemics, they generously took on the additional work required to develop the TRUST Code Supplement. While the main research and drafting burdens were shouldered by the authors of this book, all team members contributed to the consultation process. This comprised almost 60 people from 16 organisations around the world.

We are particularly grateful to our advisers, Dr. Raffaella Ravinetto, Dr Fritz Schmuhl, Emma Law, Dr. Michael Makanga, and Dr. João Monteiro, whose expertise and thoughtful feedback helped to shape the TRUST Code Supplement.

Once again, we were privileged to secure the services of one of the world's best academic editors, Paul Wise from South Africa. Paul, don't retire before we do!

This book is published in the highly successful SpringerBriefs in Research and Innovation Governance series. Our thanks go to Fritz Schmuhl for founding the series many years ago and to Doris Bleier for her excellent support from proposal

[1] The views and opinions expressed are, however, those of the authors only and do not necessarily reflect those of the European Union, the European Research Executive Agency, UKRI or SERI. Neither the European Union, the granting authority, UKRI nor SERI can be held responsible for them.

through to publication. We also thank our highly efficient and friendly production editor, Premalakshmi Gunasekaran, for all her efforts and for bearing with us during a delay.

Last but not least, we thank two anonymous reviewers whom we would have liked to thank by name for their very valuable and uplifting comments.

Following in the footsteps of its big sister, the TRUST Code, we hope this book demonstrates how the TRUST Code Supplement is grounded in real-world evidence and thus has genuine pragmatic value. Above all, we hope that the code contributes, even in a small way, to making research in fragile settings more ethical, more equitable and more firmly guided by fairness, respect, care and honesty.

Competing Interests The authors have no competing interests to declare that are relevant to the content of this manuscript.

Contents

1 The Need for a Global Code of Conduct for Research in Fragile Settings ... 1
1.1 Introduction ... 1
References ... 5

2 Why the TRUST Code Supplement Is Urgently Needed: An Illustration from Research Undertaken in Conflict Zones and Among Forcibly Displaced Populations ... 7
2.1 Introduction ... 7
2.1.1 Syria: Conflict and Humanitarian Crises ... 9
2.1.2 Syrian Asylum Seekers and Refugees in Germany ... 10
2.2 Case 1: Ethical Dilemmas in Politically Sensitive Research ... 11
2.3 Case 2: Syria—Ethical Oversight in the Absence of a Local Ethics Committee ... 13
2.4 Case 3: Publishing Research in Fragile Settings Without Local Ethics Approval ... 14
2.5 Toward Equitable and Practical Ethics in Fragile Contexts ... 15
2.6 Conclusion ... 16
References ... 16

3 Research Ethics and Integrity Challenges in Fragile Settings: A Focused Scoping Review of Academic Literature Published in English ... 21
3.1 Introduction ... 21
3.2 Methods ... 23
3.2.1 Method Selection ... 23
3.2.2 Step 1: Identifying the Research Question ... 23
3.2.3 Step 2: Identifying Relevant Studies ... 24
3.2.4 Step 3: Study Selection ... 25

3.2.5 Step 4: Charting the Data ... 28
3.2.6 Step 5: Collating, Summarising and Reporting the Results ... 28
3.3 Findings ... 29
3.3.1 Challenges for Research Participants ... 29
3.3.2 Challenges for Researchers ... 32
3.3.3 Broader Challenges for Research Ethics ... 35
3.3.4 Challenges for Research Quality and Research Integrity ... 36
3.4 Discussion ... 37
3.5 Study Limitations ... 39
3.6 Conclusion ... 39
Appendix: Literature Included in the Focused Scoping Review ... 40
References ... 45

4 Research Ethics and Integrity Challenges in Fragile Settings: A Scoping Review–Informed Evidence Mapping of German Sources ... 51
4.1 Introduction ... 51
4.2 Methods ... 52
4.2.1 Method Selection ... 52
4.2.2 Step 1: Identifying the Research Question ... 53
4.2.3 Step 2: Identifying Relevant Studies ... 54
4.2.4 Step 3: Selecting Studies to Be Included in the Review ... 55
4.2.5 Step 4: Charting the Data ... 56
4.2.6 Step 5: Collating, Summarising and Reporting the Results ... 57
4.2.7 Step 6: Consulting Stakeholders (Optional) ... 66
4.3 Discussion ... 67
4.4 Potential Shortcomings of the Review and Their Mitigation ... 68
4.5 Conclusion ... 69
References ... 70

5 Ethics and Integrity Challenges in Fragile Settings: A Survey of Experienced Researchers ... 73
5.1 Introduction ... 73
5.2 Constructing the Questionnaire ... 74
5.3 Ethical Considerations ... 76
5.4 Recruitment ... 77
5.5 Analysis ... 77
5.6 Results ... 78
5.6.1 Settings Covered ... 78
5.6.2 Positionality in Relation to the Research Setting ... 79
5.6.3 Research Ethics Approval ... 79
5.6.4 Impacts on Research Participants ... 80
5.6.5 Impacts on Local Communities ... 83

5.6.6 Impacts on Research ... 83
5.6.7 Impacts on Researchers ... 88
5.6.8 Misperception Challenges ... 88
5.6.9 Access to Research Participants ... 91
5.6.10 Other Ethics and Integrity Challenges ... 92
5.7 Interpreting the Findings ... 93
5.7.1 From Challenges to Risks to an Ethics Code ... 94
5.7.2 Reliability of the Results ... 94
5.8 Conclusion ... 95
References ... 96

6 The TRUST CODE SUPPLEMENT—A Global Code of Conduct for Research in Fragile Settings ... 99

7 Ethical Research in Fragile Settings: A Conclusion ... 105
7.1 Introduction ... 105
7.2 A Kiribati Case of Responsible Research in a Fragile Setting ... 107
7.3 Conclusion ... 114
References ... 114

Glossary ... 117

About the Authors

Doris Schroeder is Director of the Centre for Professional Ethics at the University of Lancashire UK and Professor of Moral Philosophy at the School of Law, UCLan Cyprus. She is the Lead Author of the TRUST Code and the TRUST Code Supplement.

Kate Chatfield is Reader at the Centre for Professional Ethics at the University of Lancashire UK and former editor of SAGE Research Ethics (2020–2024). She is an empirical ethicist and specialises in research ethics and global justice.

Saleh Aljadeeah is a Research Fellow at the Institute of Tropical Medicine, Antwerp. A pharmacist and global health researcher, his main research focus is access to medicines, including in fragile settings.

Hazel Partington is Senior Research Fellow in Ethics and Health at the Centre for Professional Ethics, University of Lancashire UK. She is a social scientist who moved into the field of research ethics from educational research.

Michelle Singh is a biomedical scientist at the European and Developing Countries Clinical Trials Partnership (EDCTP) Association, South Africa. She holds a doctorate in obstetrics and gynaecology, a master's in biomedical science and a master's in global bioethics.

Orla Drummond is a human rights lawyer and Senior Research Analyst with Trilateral Research in Ireland. Her main research focus is on human rights, equality and access to justice.

Young Su Park an anthropologist and medical doctor, is Assistant Professor in Medical Humanities, Seoul National University. His research focuses on global health, access to healthcare for undocumented migrants and social exclusion in health.

Lisa Tambornino is Senior Project Manager and Team Leader at the European Network of Research Ethics Committees Office in Germany. Her background is in philosophy, psychology and law, and she works primarily in research ethics and integrity.

Abbreviations

CIOMS	Council for International Organizations of Medical Sciences
DNDi	Drugs for Neglected Diseases initiative
EC	European Commission
EDCTP	European and Developing Countries Clinical Trials Partnership
EDCTP3	Global Health EDCTP
ERC	European Research Council
ERCEA	ERC Executive Agency
GCEE	German Council of Economic Experts
GDPR	General Data Protection Regulation (EU)
HIC	High-income country
LMICs	Low- and middle-income countries
NGO	Nongovernmental organisation
NRC	National Research Council (USA)
OCHA	Office for the Coordination of Humanitarian Affairs (UN)
OECD	Organisation for Economic Co-operation and Development
REC	Research ethics committee
SERI	State Secretariat for Education, Research and Innovation (Switzerland)
UKRI	UK Research and Innovation
UNHCR	United Nations High Commissioner for Refugees
WHO	World Health Organization

Chapter 1
The Need for a Global Code of Conduct for Research in Fragile Settings

Abstract As of early 2026, more than a quarter of humanity live in conflict areas and many more are affected by natural disasters, including climate change. Research has the potential to improve their situation or help them cope better with future crises. However, researchers and research participants in areas affected by conflict or disaster face significant and sometimes unique challenges. To ensure that these challenges are managed ethically and that research is conducted in line with ethical values and integrity, the TRUST Code Supplement was developed. Based on in-depth research focusing on real-world research ethics and integrity risks, the code's 23 articles guide important, but potentially risky, research. This chapter introduces a book that focuses on the reasoning for, and the research undertaken prior to drafting, the TRUST Code Supplement—A Global Code of Conduct for Research in Fragile Settings.

Keywords Research ethics · Research integrity · Fragile settings · Disaster · Conflict · TRUST Code

1.1 Introduction

Stakes and opportunities are very unevenly distributed among people around the world. Some are born into extreme poverty and die prematurely because of extreme poverty. Others are born into significant privilege and stay privileged throughout their lives. One thing they have in common is that they can all benefit from research that is undertaken ethically and that focuses on their needs.

The German Council of Economic Experts, for instance, argues that "research is fundamental for informed policy decisions" (GCEE 2023), as demonstrated during the COVID-19 pandemic and the energy crisis of the 2020s (ibid). What this council recommends is that more research data should be collected in Germany, in particular in the areas of education and wealth, to enable more effective and efficient policy decisions (ibid).

D. Schroeder et al., *Research Ethics and Integrity in Fragile Settings*,
SpringerBriefs in Research and Innovation Governance,
https://doi.org/10.1007/978-3-032-36141-7_1

An example from Kenya shows that research on the impact of climate change is particularly important. "Kenya is highly susceptible to climate change due to its varied topography, diverse climatic zones, and reliance on natural resources" (Gerard et al. 2025). Without this research, gaps in malnutrition and foodborne disease related to climate change will not be detected, which could lead to excess child deaths, as in the Horn of Africa drought in 2022 (ibid).

This book does not discuss the elephant in the room, namely that the availability of research *funding* is very unevenly distributed. It discusses the problem of undertaking research *ethically* and with *integrity*, presuming that the funding is available.

While research is likely to be important in all settings, it is more difficult to undertake in some settings compared to others. This is particularly true for fragile settings.

The OECD (2022) defines fragile settings as environments where significant risks intersect with an insufficient coping capacity to manage or mitigate those risks. This situation can lead to "negative outcomes, including violence, poverty, inequality, displacement, and environmental and political degradation" (ibid).

World Vision, a global relief, development and advocacy organisation, defines fragile settings as: "The world's most dangerous places … fraught with chronic instability, conflict and violence, trapping large numbers of people in a cycle of desperation and poverty" (Ramos 2025).

This book is about the ethics and integrity challenges for research in fragile settings. It culminates in an ethics code that can guide researchers who work in disaster-affected areas, conflict or post-conflict zones, and high-crime informal settlements, including refugee camps in lower-income settings.

The TRUST Code Supplement was built using the same methodology as the TRUST Code (TRUST 2018; Schroeder et al. 2019). This means that it was built around real-world ethics and integrity risks established through in-depth research, and then consolidated through consultations. It was not built on the drafters' views or developed from other ethics codes.

Figure 1.1 represents the code's development graphically.

The TRUST Code—A Global Code of Conduct for Equitable Research Partnerships (TRUST 2018) guides researchers who want to build equitable research partnerships. This can be particularly challenging across power differentials in international collaborations. For instance, the risk of ethics dumping (Schroeder et al. 2018) means that researchers from high-income settings might export unethical research practices that are prohibited or severely restricted in their own country to lower-income settings with possibly less regulated research environments.

While the TRUST Code can minimise the risk for ethics dumping, fragile settings pose additional challenges. For instance, one mechanism that can significantly reduce the risk of ethics dumping in international collaborations, double ethics review, might not be available in fragile settings. Researchers from higher-income countries are usually able to obtain ethics approval in their own country for research they undertake abroad, but it is not always possible to obtain ethics approval in the host country, due to, for instance, damaged governance structures after disaster or conflict. Likewise, it can be a major challenge in fragile settings to set up local complaints procedures,

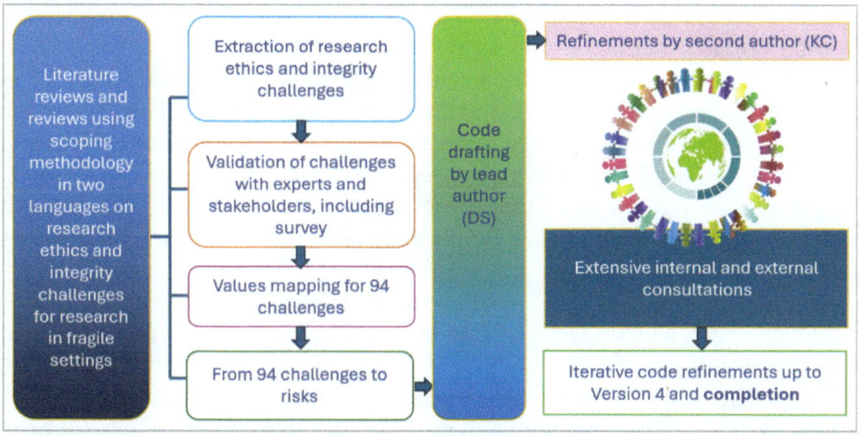

Fig. 1.1 Developing the TRUST Code Supplement

involve local communities *throughout* the research process or feed results back to research participants, all of which are requirements of the TRUST Code. While the TRUST Code remains the benchmark for equitable research partnerships, the TRUST Code Supplement provides additional guidance for research in fragile settings.

The TRUST Code Supplement is the fourth code in the TRUST Code family, depicted in Fig. 1.2.

Fig. 1.2 The TRUST Code family

Some of the adopters of the TRUST Code—including, as research funders, the European Commission, the European Research Council, the European and Developing Countries Clinical Trials Partnership (EDCTP) and Global Health EDCTP3—noted at a conference in Amsterdam in May 2024 (Gualtieri et al. 2024) that they urgently needed additional ethics guidance for fragile settings. For instance, Dr. Filipa Ferraz de Oliveira, the Ethics Principal Agent of the European Research Council (ERC) Executive Agency concluded her conference presentation in Amsterdam by saying that "the ERC would use an Ethics Code on Fragile Settings … from Day 1" (ibid: 32).

After the conference in Amsterdam, it took 16 months to develop the TRUST Code Supplement, and the results are summarised in this book in chapters that can be read independently.

Chapter 2 illustrates the difficulties of research in fragile settings using case studies from the Syrian conflict, both within Syria and among Syrian refugees in Germany. It shows how a lack of ethics guidance can hinder important research on access to medicines and healthcare, and how the TRUST Code Supplement helps address recurring ethical dilemmas and promotes approaches that are responsive to the realities of fragile settings. It further demonstrates how the code's articles can strengthen ethical reflection among researchers, institutions and also editors to ensure that affected populations are not excluded from the benefits of research.

Chapters 3 and 4 present the results of a focused scoping review (English) and a scoping review–informed evidence mapping (German). They describe and analyse the ethics and integrity risks researchers can encounter in disaster-affected areas, conflict or post-conflict zones, and high-crime informal settlements. From special burdens research participants face *before* the research even starts, to health, well-being and safety concerns for researchers, research participants and support teams, to unrealistic expectations of research participants and the potential for retraumatisation, all risks are listed and analysed. The challenges are then grouped and ultimately addressed in the TRUST Code Supplement to ensure real-world ethics guidance.

Chapter 5 validates results from literature reviews through a survey. Thirty-six researchers with first-hand experience in fragile contexts completed an online questionnaire, which comprised both open and closed questions. As a result, the challenges identified through reviews were both validated and expanded upon, leading to an evidence-informed foundation for the TRUST Code Supplement.

Chapter 6 presents the full text of the TRUST Code Supplement. Chapter 7 summarises the lessons learned from the drafting of the code and provides a good practice vignette.

Table 1.1 offers an overview of the structure of the book.

The team which drafted the TRUST Code Supplement was also involved in drafting the PREPARED Code—A Global Code of Conduct for Research During Pandemics (PREPARED 2025; see also Chatfield and Singh 2025). While it is reasonable to hope that the PREPARED Code will never have to be put into practice, there can be *no* hope that this TRUST Code Supplement will not be needed.

In his New Year message for 2026, UN Secretary-General António Guterres warns that "conflict, climate breakdown and systemic violations of international law are

Table 1.1 Structure of the book

Chapter	Content	Type of chapter
1	Overview of book	Introduction
2	Rationale of book explained through a Syrian case study	Case study
3	Research ethics and integrity challenges for research in fragile settings published in English (PubMed)	Research
4	Research ethics and integrity challenges for research in fragile settings published in German (Google Scholar)	Research
5	Survey of researchers with experience of work in fragile settings on the research ethics and integrity challenges of their work	Research
6	The TRUST Code Supplement—A Global Code of Conduct for Research in Fragile Settings	Ethics guidance
7	Lessons learned from research with one case vignette	Conclusion with one case study

eroding trust in leadership worldwide" and that "more than a quarter of humanity now lives in conflict-affected areas" (UN 2025).

With research-based, real-world guidance on research ethics and integrity risks in fragile settings, it might at least be possible to ensure that life-saving, important research is undertaken ethically.

References

Chatfield K, Singh M (2025) Research ethics and integrity during pandemics: not unique, but vastly magnified challenges. In: Chatfield K, Singh M (eds) Research ethics and integrity during pandemics: developing the PREPARED code. SpringerBriefs in research and innovation governance. Springer, Cham, pp 1–7. https://doi.org/10.1007/978-3-031-91324-2

GCEE (2023) Germany needs a research data infrastructure for the 21st century. Press release. German Council of Scientific Experts. https://www.sachverstaendigenrat-wirtschaft.de/fileadmin/dateiablage/gutachten/jg202324/JG202324_PressRelease_Chapter_6.pdf. Accessed 26 Jan 2026

Gerard J, Kibaara T, Blom IM et al (2025) Climate-sensitive health outcomes in Kenya: a scoping review of environmental exposures and health outcomes research, 2000–2024. Climate 13(7):133. https://doi.org/10.3390/cli13070133

Gualtieri J, Kornioti N, Schroeder D (2024) PREPARED Amsterdam conference report. https://prepared-project.eu/wp-content/uploads/2025/08/PREPARED-Amsterdam-report_compressed.pdf. Accessed 26 Jan 2026

OECD (2022) How fragile contexts affect the well-being and potential of women and girls. OECD Development Co-operation Directorate. OECD Publishing, Paris. https://www.oecd.org/en/publications/how-fragile-contexts-affect-the-well-being-and-potential-of-women-and-girls_c8b2287c-en.html. Accessed 29 Dec 2025

PREPARED (2025) The PREPARED code: a global code of conduct for research during pandemics. https://prepared-project.eu/prepared-code/. Accessed 10 Sept 2025

Ramos M (2025) Fragile contexts: the world's most dangerous places. 18 August. World Vision Canada. https://www.worldvision.ca/en/stories/what-are-fragile-contexts. Accessed 29 Dec 2025

Schroeder D, Cook J, Hirsch F, Fenet S, Muthuswamyet V (eds) (2018) Ethics dumping: case studies from north-south research collaborations. Springer, Cham. https://doi.org/10.1007/978-3-319-64731-9

Schroeder D, Chatfield K, Singh M et al (2019) Equitable research partnerships: a global code of conduct to counter ethics dumping. Springer Nature, Cham. https://doi.org/10.1007/978-3-030-15745-6

TRUST (2018) The TRUST Code: a global code of conduct for equitable research partnerships. https://doi.org/10.48508/GCC/2018.05

UN (2025) UN chief urges world leaders to 'get priorities straight' as New Year message calls for peace over war. UN News, 29 December 2025. https://news.un.org/en/story/2025/12/1166677. Accessed 28 Jan 2026

Chapter 2
Why the TRUST Code Supplement Is Urgently Needed: An Illustration from Research Undertaken in Conflict Zones and Among Forcibly Displaced Populations

Abstract Armed conflicts and forced displacement have reached unprecedented levels, creating complex health and humanitarian challenges. Conducting research in such fragile contexts is both necessary and ethically demanding, as standard oversight mechanisms may be absent or not fully adapted to local realities. This chapter illustrates how the TRUST Code Supplement (PREPARED 2025) provides practical and context-sensitive guidance for conducting ethical research in crisis-affected environments. Drawing on case studies from research on access to medicines and healthcare among populations affected by the Syrian conflict, both within Syria and among refugees in Germany, the chapter shows how the TRUST Code Supplement helps address recurring ethical dilemmas and promotes approaches that are responsive to the realities of fragile settings. It further demonstrates how the code's articles can strengthen ethical reflection among researchers, institutions and editors to ensure that affected populations are not excluded from the benefits of research. Yet, while it marks significant progress, it should also be seen as a starting point rather than an endpoint, as fragile contexts are constantly evolving and continued dialogue and refinement of ethical frameworks remain essential to keep pace with these realities.

Keywords Conflict · Forced migration · Fragile contexts · Research ethics · TRUST Code Supplement

2.1 Introduction

Armed conflicts are escalating to levels unseen since World War II, with 2024 recording the highest number of state-based conflicts in more than seven decades and being one of the worst years in modern history for civilian populations caught in conflict (OCHA 2024; PRIO 2025).

The United Nations Office for the Coordination of Humanitarian Affairs (OCHA) projected in December 2024 that 305 million people worldwide would require humanitarian assistance and protection in 2025 (OCHA 2024). Updated figures in mid-2025 suggest that needs are already closely matching or even exceeding that

D. Schroeder et al., *Research Ethics and Integrity in Fragile Settings*,
SpringerBriefs in Research and Innovation Governance,
https://doi.org/10.1007/978-3-032-36141-7_2

projection, with over 299 million individuals identified as people in need in global response plans (ReliefWeb 2025).

The Southern and Eastern Africa regions host the largest number of people in need of humanitarian aid, followed closely by the Middle East and North Africa, where the Syrian conflict and the worsening situation in the Occupied Palestinian Territory and Lebanon are driving unprecedented levels of vulnerability. At the same time, protracted conflicts and other crises in Afghanistan, Myanmar, Venezuela and Ukraine demonstrate that the global scale of displacement and humanitarian emergencies is not only increasing but also becoming more complex and prolonged (OCHA 2024).

Armed conflicts increase and change the healthcare needs of the populations caught in conflict zones. They often increase the prevalence of penetrating traumatic injuries with contaminated open wounds and fractures (Abbara et al. 2018). Additionally, the interruption of vaccination campaigns contributes to the resurgence of preventable diseases; the collapse of health systems leads to gaps in primary care and the loss of specialised services; and deteriorating water and sanitation conditions increase the burden of communicable diseases. At the same time, shortages of medicines and medical supplies undermine the management of non-communicable diseases (Leyh and Gispen 2018; Aljadeeah et al. 2025). Together, these factors create complex and multilayered health crises that demand evidence-based responses.

Conducting research in conflict zones is fraught with practical and ethical challenges. Access to study populations is often restricted by security and safety risks, which can endanger both research participants and researchers (Kohrt et al. 2019). Even when access is possible, research is undermined by methodological barriers such as the absence of reliable population denominators, and by the high mobility of populations and individuals, which limits the possibilities for the follow-up that is crucial for some research study designs (Abouzeid et al. 2021). Furthermore, research activities themselves may unintentionally cause harm. For example, asking participants to recall traumatic experiences can exacerbate distress, or interviewing overburdened healthcare providers could divert their already limited time away from patients, thereby undermining the delivery of essential healthcare services (Ford et al. 2009).

Yet despite these challenges, the justification for conducting research in conflict-affected settings is compelling. Populations caught in crises have the same right to healthcare and life-saving interventions as those living in stable contexts. One could argue that health professionals and humanitarian organisations have an ethical duty to ensure that their interventions are evidence-informed and relevant to the needs of affected communities (Ford et al. 2009).

Research provides crucial means of evaluating the feasibility, safety and efficiency of interventions in these fragile settings. Excluding displaced or conflict-affected populations from scientific inquiry risks compounding their marginalisation. In fact, the systematic exclusion of such groups can itself be viewed as unethical, as it denies them the benefits of knowledge production that could improve their access to healthcare and protection (Ford et al. 2009).

With the record-breaking numbers of people affected by conflicts globally, whether living in conflict zones or forcibly displaced, it becomes crucial to reflect on how research can be conducted ethically in such challenging contexts.

In this chapter we draw on the research experiences of one of our co-authors to illustrate these dilemmas through three case studies involving populations affected by the Syrian conflict: communities living in conflict-affected northern Syria and Syrian asylum seekers and refugees in Germany. Background on Syria precedes the case studies.

The case studies reveal the complex ethical challenges that arise in both contexts, ranging from navigating fragile infrastructures and security risks to addressing the vulnerabilities and rights of displaced populations in high-income settings. They also highlight the relevance of the TRUST Code Supplement (PREPARED 2025) as a framework for promoting fairness, respect, care and honesty in research conducted in fragile and crisis-affected environments.

2.1.1 Syria: Conflict and Humanitarian Crises

Syria is a country in the eastern Mediterranean region that experienced a prolonged armed conflict between 2011 and 2024. This armed conflict caused one of the largest humanitarian crises since the Second World War (UN News 2025). Before 2011, the Syrian health system resembled that of other middle-income countries, with notable improvements in health indicators such as reduced infant and maternal mortality and rising life expectancy. The country has also undergone an epidemiological transition from infectious diseases to non-communicable diseases over the three decades since the 1980s (WHO 2012; Aljadeeah et al. 2021; Ansbro et al. 2021).

The conflict that began in 2011 displaced millions of Syrians, with 6.8 million people forced to leave the country and more than 6 million internally displaced (Devi 2021). The conflict had a devastating impact on the health system in Syria. Health workers and facilities became targets of repeated attacks: more than half of the country's health facilities were destroyed or damaged, and over two-thirds of health professionals left the country as a consequence of the escalation of violence (Bdaiwi et al. 2020; ICRC 2021). What remained of the system became fragmented, with different authorities controlling health services in areas held by the Al-Assad government, the Autonomous Administration of North and East Syria, opposition-held territories in the northwest, and Turkish-controlled areas along the border with Türkiye (Abbara et al. 2021). Humanitarian organisations stepped in to fill some of the gaps, but needs remained acute. The conflict also shaped the landscape of health research on Syria. Approximately 72% of health-related publications between 1991 and 2017 were published after the war began, with increasing focus on conflict-related trauma, mental health and injuries (Abdul-Khalek et al. 2020).

On 8 December 2024, the Al-Assad regime came to an end after more than five decades of authoritarian rule characterised by systematic repression, human rights abuses and crimes against humanity (Global Centre for the Responsibility to Protect

2024). Despite the fall of the regime, Syria remains fragile. Reports of mass civilian killings on the northwest coast and the 2025 clashes in Suwayda Governorate, which included civilian deaths, kidnappings and the destruction of property, illustrate the ongoing insecurity. Both Amnesty International and the Office of the UN High Commissioner for Human Rights have expressed alarm and called for accountability and the protection of civilians (Amnesty International 2025; OHCHR 2025).

2.1.2 Syrian Asylum Seekers and Refugees in Germany

According to the UNHCR, there were more than 123.2 million displaced people globally by the end of 2024, with Syrians representing the largest refugee population at 6 million, followed by Venezuelans, Afghans and South Sudanese (UNHCR 2024). Most displaced people remain in countries bordering their homelands, with the Islamic Republic of Iran hosting 3.5 million refugees and Türkiye 2.9 million. These were followed by Colombia with 2.8 million and Germany with 2.7 million refugees by the end of 2024 (UNHCR 2024).

Germany, a signatory to the 1951 Geneva Refugee Convention since 1953, grants refugees the right to access healthcare and other services (UNHCR 2021–2025). Historically, asylum seekers also had full access to the ordinary healthcare system, but this changed in 1993 with the introduction of the Asylum Seekers Benefits Act (Asylbewerberleistungsgesetz, or AsylbLG) (Bundesamt für Justiz 1993; Jaschke and Kosyakova 2021). Under this legislation, asylum seekers became entitled to only a limited set of healthcare services: emergency treatment, management of acute and painful conditions, maternity care, vaccinations and selected preventive services (Bundesamt für Justiz 1993). Additional care is possible, but subject to administrative approval.

This restriction has been heavily criticised for its human rights implications. International bodies, including the UN Committee on Economic, Social and Cultural Rights, have called on Germany to revise the AsylbLG and ensure equal access to curative, preventive and palliative health services irrespective of legal status (UN Economic and Social Council 2018). Scholars have likewise argued that differentiating healthcare access based on residency status violates the fundamental right to health (Bozorgmehr et al. 2017). Overall, while Germany remains one of the leading host countries for asylum seekers, significant challenges persist in ensuring equitable access to healthcare. The situation of asylum seekers in Germany illustrates the broader tension between international commitments to protecting forcibly displaced populations and the realities of national policy frameworks, and underlines the urgent need for highlighting the importance of ethically grounded research to guide context-sensitive approaches in health and humanitarian governance.

The three case studies below concern research conducted by book co-author **Dr Saleh Aljadeeah**, and are therefore presented by him in the first person.

2.2 Case 1: Ethical Dilemmas in Politically Sensitive Research

In my previous research on access to medicines among asylum seekers and refugees in Germany, one study was delayed for about 6 months by difficulties in obtaining approval from the relevant research ethics committee (REC). I offer this example as a contribution to a shared learning process about research ethics review systems in politically sensitive contexts, and about the importance of ethics guidelines focused on the ethical concerns of conducting research in fragile settings, such as the TRUST Code Supplement (PREPARED 2025), in addressing these challenges and concerns.

Germany hosted the highest number of asylum seekers in Europe between 2015 and 2022 (IOM 2024), and the topic of migration has been, and remains, at the centre of political debate in that country (Deutschlandfunk 2025). In such a highly politicised environment, any research that does not carefully address ethical concerns—by, for instance, including robust data protection measures for asylum seekers and refugees—risks causing serious harm. As the United Nations High Commissioner for Refugees (UNHCR) has emphasised, data protection is part and parcel of refugee protection: refugees are often in a vulnerable position, and their information is highly sensitive. If misused or accessed by unauthorised actors, such data could expose refugees to risks far greater than those faced by the general population (UNHCR 2018). For this reason, I was glad that my institution's REC recognised the sensitivity of my topic and invested significant time and effort in reviewing my study protocol.

On the other hand, the lack of contextual expertise in the REC regarding research with refugees and asylum seekers created unnecessary obstacles during the review process, in my view. One concern raised by the REC was that my sample would not be randomly drawn, and therefore might not be representative. In my response, I explained that a purely random sample was not feasible, since there was no sampling frame covering the asylum seeker and refugee population in Germany for researchers to use (Aljadeeah 2022), and that I had therefore combined several sampling approaches in order to achieve the best representativeness possible under the circumstances (Aljadeeah 2022).

Another concern was that standard demographic questions, such as those concerning age, sex or level of education, were considered too risky, as they might make participants identifiable. To address this, I revised the survey design to strengthen anonymity (e.g., by asking participants to state what ten-year age group they fell into, rather than their exact age). Despite these adjustments, the committee still believed that the research could cause harm due to data protection issues and initially withheld approval. At that point, I sought advice from the data protection officer of my university, who reviewed my protocol and issued a statement confirming that my study complied with the university's data protection regulations. On that basis, the REC eventually granted approval, but only after significant delays.

Existing ethical guidelines, such as the Declaration of Helsinki on medical research involving human participants (WMA 2024), have long been invaluable tools for reviewing research protocols. However, in fragile settings, including contexts of forced displacement, RECs are often confronted with dilemmas that fall outside the scope of standard guidance. They may face heightened concerns over participant safety, data security, community acceptance and the risks of unintended harm, yet they lack concrete, context-specific strategies for navigating these issues. This is precisely where the TRUST Code Supplement can serve as a practical compass. It does not replace the responsibility of RECs to critically review research proposals, but it does offer structured, context-sensitive pathways for addressing ethical concerns without unnecessarily delaying vital research.

For example, Article 5 addresses a recurring challenge: RECs may, when reviewing studies in fragile contexts, fear that approval could expose participants or researchers to harm. The TRUST Code Supplement encourages RECs to engage supportively with researchers to find solutions that make the research safer and more ethically sound, rather than simply halt or indefinitely postpone it. This approach reframes the review process into a constructive dialogue aimed at enabling ethically robust research.

Similarly, Article 7 acknowledges that unstable environments—a common characteristic of fragile settings, where shifts in the situation, changes in political power, and the emergence of new challenges and risks are frequent—can require protocol adaptations during the course of a study. RECs might be concerned that such adjustments could compromise research integrity or deviate from approved procedures. The TRUST Code Supplement guides committees in assessing whether proposed changes are locally appropriate, feasible and still aligned with the study's core ethical commitments, a balance that allows the necessary flexibility while maintaining scientific rigour.

Article 11 highlights another frequent concern, namely whether the research design adequately reflects local realities and cultural nuances. RECs may hesitate to approve studies that appear externally driven or disconnected from the communities involved. By urging the integration of local knowledge and perspectives, the TRUST Code Supplement provides a constructive way forward, encouraging RECs to look for evidence of meaningful local engagement, such as engagement with local advisory boards, community consultations or partnerships with local researchers.

As another example, Articles 13 and 14 of the TRUST Code Supplement address the risk that research agendas may be shaped solely by external actors, sidelining the voices and priorities of affected communities in fragile settings. RECs can use these articles as a benchmark for assessing whether local stakeholders have been involved in decision-making processes and whether their local knowledge was gathered. Such involvement not only strengthens the ethical foundation of the research but also gives RECs the additional assurance that the study is locally relevant and acceptable, and can be responsibly implemented.

In this way, the TRUST Code Supplement offers more than abstract ethical principles; it provides RECs with concrete considerations and constructive suggestions to navigate their decisions. By framing concerns alongside practical solutions, the code helps committees uphold rigorous ethical standards while avoiding the unintended consequence of obstructing much-needed research in fragile settings.

2.3 Case 2: Syria—Ethical Oversight in the Absence of a Local Ethics Committee

I refer here to a research project I conducted in 2023–2024 on access to medicines for noncommunicable diseases in conflict-affected northern Syria. The standard practice of securing local research ethics approval, as, for instance, recommended by the TRUST Code (TRUST 2018), the big sister of the TRUST Code Supplement, was not feasible in such a fragile context. Local RECs were either entirely absent or operating under authorities whose involvement could have posed significant risks to participants' safety and data privacy. The absence of a local REC, or of one that can be trusted to adhere to recognised ethical standards and keep conflicts of interest to the minimum, is common in fragile settings. This situation, however, should not be allowed to block research. Even when local ethics review is not possible, the overarching ethical question remains as relevant in fragile settings as in any other context: how can research be conducted in a way that it benefits societies and individuals, while minimising potential harm to those very societies and individuals, and indeed to the researchers themselves? Addressing this question requires careful consideration of local realities, power dynamics and potential risks.

Given these realities in northern Syria, I sought research ethics approval from an international REC with proven experience in reviewing studies conducted in fragile settings. While this met the formal requirement of a research ethics review, I was also aware of the need to ground the research in the lived realities of the affected population. To achieve this, a local research advisory group was established, composed of patients, healthcare professionals and members of the Syrian diaspora. In the absence of functioning local institutional research ethics review structures, this group played a crucial role in ensuring the relevance of the study, assessing and enhancing its feasibility, and safeguarding its ethical rigour.

Designing and implementing research that is of genuine benefit to societies and individuals, while minimising potential harm to them and to researchers, requires practical, context-sensitive guidance, especially when conventional oversight mechanisms are absent. The TRUST Code Supplement provides clear and actionable direction for navigating such circumstances. Article 10 explicitly addresses situations where local ethics review within the host country is not possible, requiring researchers to justify this to the approving committee and to explain what alternative ethical safeguards will be put in place. Additionally, Article 14 calls for local stakeholders to be involved in decision-making processes to ensure that research priorities and implementation strategies are not driven solely by external actors and to avoid cultural misunderstandings. Furthermore, Article 13 underscores the value of integrating local knowledge and community perspectives into research design and implementation, a principle directly applied through the establishment of the advisory group.

By following this guidance, researchers and RECs can work together to design ethically sound studies even when conventional oversight mechanisms, such as double ethics review, are absent. The TRUST Code Supplement not only acknowledges the structural limitations common in fragile settings but also offers constructive suggestions, such as extending insurance cover to local researchers, and ensuring that research remains relevant and respectful to the communities it seeks to serve.

2.4 Case 3: Publishing Research in Fragile Settings Without Local Ethics Approval

Academic journals, including those on global and public health, usually require proof of local ethics approval as a prerequisite for publishing research involving human participants. This policy reflects a strong and commendable commitment to ethical research practices. However, as demonstrated in Case 2, securing such approval in fragile settings such as conflict-affected areas or post-disaster zones is often impossible. In these contexts, researchers often find themselves in the position of having to justify their work post hoc to editorial boards. This can be difficult, not because the research was conducted without ethical safeguards, but because, until recently, there was no widely recognised framework for explaining why local ethics review was not feasible and what alternative measures were taken to protect participants and uphold ethical standards. Without such a framework, these justifications can seem ad hoc, potentially undermining confidence in the research and delaying its dissemination.

The TRUST Code Supplement offers a structured, credible and context-sensitive reference for addressing this challenge. Article 10 provides advice for situations where local ethics review is not possible, guiding researchers in the first instance to explain this to the approving REC, and possibly later also to any editorial teams. Just as researchers who follow the TRUST Code Supplement can justify to the approving REC why they could not obtain local ethics review approval, they can describe to editors the alternative safeguards they implemented. Importantly, they can also point to their discussions with the approving REC on the topic.

Article 14 calls for local stakeholders to be involved in research decision-making, to ensure that the absence of a formal local ethics review does not result in the exclusion of local voices. Article 13 underscores the importance of integrating local knowledge and perspectives, something that can be demonstrated through meaningful community engagement, such as establishing advisory groups. Additionally, Article 19 urges researchers to provide feedback to participants, and thus demonstrate that research can remain respectful and inclusive despite the absence of conventional oversight mechanisms.

Beyond supporting researchers, the TRUST Code Supplement can also serve as a practical guide for journal editors when they are evaluating manuscripts that have emerged from fragile settings. It provides an established ethical reference that can help editors assess whether the alternative safeguards employed were appropriate, and whether the research met the same ethical principles expected in more stable environments. This would align with Springer Nature's policy of asking questions aligned with the original TRUST Code (2018) when they assess submissions that pose risks of ethics dumping (Schroeder et al. 2018) or helicopter research (Nature 2022).

The values in the TRUST Code Supplement could also inform updates to widely used research reporting guidelines hosted by the EQUATOR Network, such as STROBE (for observational studies) and STARD (for diagnostic accuracy studies), These checklists focus primarily on methodological transparency and completeness, but they generally do not require authors to report on research ethics approval, informed consent or alternative research ethics safeguards. This gap is particularly relevant to research in fragile settings, where conventional local ethics review may not be possible.

Drawing on the TRUST Code Supplement, new checklist items could prompt authors to explain how ethics oversight was handled in such contexts and what measures were taken to protect participants when standard local ethics approval mechanisms were absent or compromised. This would not only improve transparency but also help journal editors and reviewers evaluate whether ethical principles were upheld in line with the realities of fragile settings.

By embedding these considerations into both editorial practice and reporting standards, the TRUST Code Supplement can help ensure that research from fragile settings is reported and evaluated in a way that safeguards participants, supports transparency and enables valuable findings to reach the public domain without compromising ethical integrity.

2.5 Toward Equitable and Practical Ethics in Fragile Contexts

For the TRUST Code Supplement to fulfil its potential, it must move beyond being a useful tool for individual researchers and become a widely recognised reference point for research ethics review processes in fragile and crisis-affected settings. Institutionalising its use would not only create greater consistency across research projects but also provide research ethics committees with a structured framework to navigate dilemmas that conventional guidelines rarely address.

Capacity building within research ethics committees is equally important. Committees may lack the contextual knowledge or expertise required to evaluate research in humanitarian and conflict-affected environments. Investing in training and in the exchange of experience between committees in stable settings and those in fragile circumstances, as well as in researchers with significant experience of working in fragile settings, could strengthen their ability to provide timely, constructive and context-sensitive feedback. This, in turn, would allow research ethics review to act as an enabler for urgent research.

Finally, sustainable, locally grounded mechanisms are needed to fill the gaps where institutional review structures are absent or compromised. In health research, advisory groups made up of patients, community members, healthcare providers and diaspora representatives offer one such mechanism. While not a replacement for institutional ethics review, they can play a vital role in ensuring that research is relevant, feasible and respectful of local realities, while also enhancing accountability to affected populations.

2.6 Conclusion

The TRUST Code Supplement represents a milestone in the global conversation on ethical research in fragile settings. It recognises that ethical research is not only about preventing harm but also about enabling knowledge generation that is respectful, relevant and responsive to local contexts. By providing practical pathways for researchers, research ethics committees and journal editors, the code helps ensure that displaced and crisis-affected populations are not excluded from the benefits of ethically conducted research.

At its core, the TRUST Code Supplement provides a critical ethical anchor: it prevents exclusionary practices and guards against the risk that ethics processes themselves can become barriers to much-needed research. Yet, while it marks significant progress, it should also be seen as a starting point rather than an endpoint. Fragile contexts are constantly evolving, shaped by shifting political, security and humanitarian landscapes. Continued dialogue, the iterative refinement of ethical frameworks and the development of context-driven concepts will be essential to keep pace with these realities. In this way, the TRUST Code Supplement both anchors and advances the global effort to conduct research that upholds fairness, respect, care and honesty, even under the most challenging circumstances.

References

Abbara A, Rawson TM, Karah N, El-Amin W, Hatcher J, Tajaldin B, Dar O, Dewachi O, Abu Sitta G, Uhlin BE, Sparrow A (2018) Antimicrobial resistance in the context of the Syrian conflict: drivers before and after the onset of conflict and key recommendations. Int J Infect Dis 73:1–6. https://doi.org/10.1016/j.ijid.2018.05.008

Abbara A, Marzouk M, Mkhallalati H (2021) Health system fragmentation and the Syrian conflict. In: Bseiso J, Hofman M, Whittall J (eds) Everybody's war: the politics of aid in the Syria crisis. Oxford University Press, pp 33–57. https://doi.org/10.1093/oso/9780197514641.003.0003

Abdul-Khalek RA, Kayyal W, Akkawi AR, Almalla M, Arif K, Bou-Karroum L, El-Harakeh A, Elzalabany MK, Fadlallah R, Ghaddar F, Kashlan D, Kassas S, Khater T, Mobayed N, Rahme D, Saifi O, Jabbour S, El-Jardali F, Akl EA, Jawad M (2020) Health-related articles on Syria before and after the start of armed conflict: a scoping review for The Lancet–American University of Beirut Commission on Syria. Confl Heal 14(1):73. https://doi.org/10.1186/s13031-020-00316-7

Abouzeid M, Elzalabany MK, Nuwayhid I, Jabbour S (2021) Conflict-related health research in Syria, 2011–2019: a scoping review for The Lancet–AUB Commission on Syria. Confl Heal 15(1):92. https://doi.org/10.1186/s13031-021-00384-3

Aljadeeah S, Tarrafeta B, Fahed S, Kielmann K, Ravinetto R (2025) Access to essential medicines for noncommunicable diseases during conflicts: the case cardiovascular diseases, diabetes and epilepsy in northern Syria. PLOS Glob Public Health 5(12):e0004744. https://doi.org/10.1371/journal.pgph.0004744

Aljadeeah S, Nagel E, Wirtz VJ (2021) Outpatient cardiovascular diseases and diabetes medicines dispensing in the population with government health insurance in Syria between 2018 and 2019: a retrospective analysis. BMC Health Serv Res 21(1):1088. https://doi.org/10.1186/s12913-021-07124-6

Aljadeeah S, Tarrafeta B, Fahed S, Kielmann K, Ravinetto R (2025) Access to essential medicines for noncommunicable diseases during conflicts: the case cardiovascular diseases, diabetes and epilepsy in northern Syria. medRxiv. https://doi.org/10.1101/2025.05.20.25327993

Amnesty International (2025) Syria: horrific killings of civilians on northwest coast must be investigated, 10 March. https://www.amnesty.org/en/latest/news/2025/03/syria-horrific-killings-of-civilians-on-northwest-coast-must-be-investigated/. Accessed 28 Sept 2025

Ansbro É, Homan T, Prieto Merino D, Jobanputra K, Qasem J, Muhammad S, Fardous T, Perel P (2021) Clinical outcomes in a primary-level non-communicable disease programme for Syrian refugees and the host population in Jordan: a cohort analysis using routine data. PLoS Med 18(1):e1003279. https://doi.org/10.1371/journal.pmed.1003279. Erratum in PLoS Medicine 21(3):e1004373. https://doi.org/10.1371/journal.pmed.1004373

Bdaiwi Y, Rayes D, Sabouni A, Murad L, Fouad F, Zakaria W, Hariri M, Ekzayez A, Tarakji A, Abbara A (2020) Challenges of providing healthcare worker education and training in protracted conflict: a focus on non-government controlled areas in North West Syria. Confl Heal 14:42. https://doi.org/10.1186/s13031-020-00287-9

Bozorgmehr K, Wenner J, Razum O (2017) Restricted access to health care for asylum-seekers: applying a human rights lens to the argument of resource constraints. Eur J Pub Health 27(4):592–593. https://doi.org/10.1093/eurpub/ckx086

Bundesamt für Justiz (1993) Asylbewerberleistungsgesetz (AsylbLG). https://www.gesetze-im-internet.de/asylblg/BJNR107410993.html. Accessed 28 Sept 2025

Deutschlandfunk (2025) Bundestagswahl 2025—Wahlprogramme: Migration. https://www.deutschlandfunk.de/bundestagswahl-2025-wahlprogramme-migration-100.html. Accessed 28 Sept 2025

Devi S (2021) Health in Syria: a decade of conflict. Lancet 397(10278):955–956. https://doi.org/10.1016/S0140-6736(21)00618-8

Ford N, Mills EJ, Zachariah R, Upshur R (2009) Ethics of conducting research in conflict settings. Confl Heal 3:7. https://doi.org/10.1186/1752-1505-3-7

Global Centre for the Responsibility to Protect (2024) Statement on the developing situation in Syria. https://www.globalr2p.org/publications/statement-on-the-developing-situation-in-syria/. Accessed 28 Sept 2025

ICRC (2021) Health-care providers, patients suffer thousands of attacks on health-care services over the past five years, ICRC data show. International Committee of the Red Cross, 3 May. https://www.icrc.org/en/document/health-care-providers-patients-suffer-thousands-attacks-health-care-services-past-5-years. Accessed 28 Sept 2025

IOM (2024) World Migration Report 2024. International Organization for Migration. https://publications.iom.int/books/world-migration-report-2024. Accessed 28 Sept 2025

Jaschke P, Kosyakova Y (2021) Does facilitated and early access to the healthcare system improve refugees' health outcomes? Evidence from a natural experiment in Germany. Int Migr Rev 55(3):812–842. https://doi.org/10.1177/0197918320980413

Kohrt BA, Mistry AS, Anand N, Beecroft B, Nuwayhid I (2019) Health research in humanitarian crises: an urgent global imperative. BMJ Glob Health 4:e001870. https://doi.org/10.1136/bmjgh-2019-001870

Leyh BM, Gispen ME (2018) Access to medicines in times of conflict: overlapping compliance and accountability frameworks for Syria. Health Hum Rights 20(1):237–250. PMID: 30008566, PMC6039728. https://www.hhrjournal.org/2018/04/09/access-to-medicines-in-times-of-conflict-overlapping-compliance-and-accountability-frameworks-for-syria/. Accessed 28 Sept 2025
Nature (2022) Nature addresses helicopter research and ethics dumping (editorial). Nature 606(7). https://doi.org/10.1038/d41586-022-01423-6
OCHA (2024) Global Humanitarian Overview 2025. United Nations Office for the Coordination of Humanitarian Affairs. https://www.unocha.org/events/global-humanitarian-overview-2025. Accessed 28 Sept 2025
OHCHR (2025) UN Syria commission deeply alarmed by Suwayda violence, urges de-escalation, civilian protection and respect for human rights. United Nations High Commissioner for Human Rights, 15 July. https://www.ohchr.org/en/press-releases/2025/07/un-syria-commission-deeply-alarmed-suwayda-violence-urges-de-escalation. Accessed 28 Sept 2025
PREPARED (2025) The Trust Code Supplement—a global code of conduct for research in fragile settings. https://www.globalcodeofconduct.org/trust-code-supplement/. Accessed 10 Oct 2025
PRIO (2025) New data shows conflict at historic high as U.S. signals retreat from world stage. Peace Research Institute Oslo, 9 June. https://www.prio.org/news/3616. Accessed 28 Sept 2025
ReliefWeb (2025) Global Humanitarian Overview. https://reliefweb.int/topics/global-humanitarian-overview. Accessed 28 Sept 2025
Schroeder D, Cook J, Hirsch F, Fenet S, Muthuswamyet V (eds) (2018) Ethics dumping: case studies from North-South research collaborations. Springer, Cham. https://doi.org/10.1007/978-3-319-64731-9
TRUST (2018) The TRUST Code—a global code of conduct for equitable research partnerships. https://doi.org/10.48508/GCC/2018.05
UN Economic and Social Council (2018) Concluding observations on the sixth periodic report of Germany 2018, 27 Nov. https://www.ohchr.org/en/documents/concluding-observations/ec12deuco6-concluding-observations-sixth-periodic-report-germany. Accessed 28 Sept 2025
UN News (2025) Syria's humanitarian crisis: 16.5 million in need amid continuing conflict. UN News Global Perspective Human Stories, 20 March. https://news.un.org/en/story/2025/03/1161326. Accessed 13 Oct 2025
UNHCR (2018) Data protection is part and parcel of refugee protection. Office of the UN High Commissioner for Refugees, 23 May. https://www.unhcr.org/blogs/data-protection-part-parcel-refugee-protection/. Accessed 28 Sept 2025
UNHCR (2021–2025): About us: Office of the UN High Commissioner for Refugees. https://www.unhcr.org/about-us.html. Accessed 28 Sept 2025
UNHCR (2024) Global trends report 2024. Office of the UN High Commissioner for Refugees. https://www.unhcr.org/global-trends-report-2024. Accessed 28 Sept 2025
WHO (2012) Noncommunicable diseases country profiles 2011. World Health Organization, Geneva. https://www.who.int/publications/i/item/9789241502283. Accessed 28 Sept 2025
WMA (2024) Declaration of Helsinki: ethical principles for medical research involving human participants. World Medical Association. https://www.wma.net/what-we-do/medical-ethics/declaration-of-helsinki/. Accessed 28 Sept 2025

Chapter 3
Research Ethics and Integrity Challenges in Fragile Settings: A Focused Scoping Review of Academic Literature Published in English

Abstract Research in fragile settings is essential to inform humanitarian, health and development responses, yet such contexts pose heightened challenges for research ethics and research integrity due to factors like instability, vulnerability and power asymmetries. To inform the development of a new ethics code for research in fragile settings, this study aimed to identify and synthesise ethics- and integrity-related challenges that arise or are exacerbated when conducting research in fragile settings. To this end, a focused scoping review was conducted, sourcing relevant English language literature from across two major databases. Challenges were grouped into four domains: challenges for research participants, challenges for researchers, broader ethics challenges, and challenges for research quality and research integrity. Key issues that were identified included compromised informed consent, risks of physical and emotional harm, exploitation, researcher insecurity, weak ethics oversight, power asymmetries and compromised methodological rigour. Overall, the reviewed literature revealed that there are wide-ranging challenges for research ethics and integrity in fragile settings, which are complex and interrelated, underscoring the need for the context-sensitive ethics guidance that the TRUST Code Supplement provides to support responsible research in crisis-affected environments.

Keywords Research ethics · Fragile settings · Disaster settings · Humanitarian settings · Research integrity · Scoping review

3.1 Introduction

Research in crisis and fragile contexts can be essential to inform public health action and ensure that humanitarian responses and recovery efforts are grounded in lived realities (CIOMS 2016; WHO 2020). Research plays a vital role not only in understanding and responding to challenges in fragile settings, but also in informing effective policymaking, programme design, monitoring and evaluation (Idris 2019). At the same time, individuals and communities in these contexts can be experiencing conflict, displacement, disaster, disease or extreme poverty; circumstances

D. Schroeder et al., *Research Ethics and Integrity in Fragile Settings*,
SpringerBriefs in Research and Innovation Governance,
https://doi.org/10.1007/978-3-032-36141-7_3

that heighten vulnerability and magnify power asymmetries (OECD 2022; Davidson et al. 2024). Research activities themselves may also entail significant risks, including physical insecurity, psychological harm, surveillance or retaliation, affecting both participants and researchers operating in unstable and unpredictable environments (Almeida and Harris 2021; Kurihara et al. 2022).

Against this backdrop, both historical and more recent accounts of ethics and integrity challenges demonstrate that unethical research practices have occurred in crisis settings, including inadequate consent processes, extractive data collection, and failures to minimise harm or ensure an equitable distribution of research benefits. Such practices have fuelled mistrust and contributed to the critique of humanitarian and health research conducted in crisis-affected contexts (Ebubedike et al. 2023; Franc and O'Mathúna 2025). They have also exposed the inadequacy of relying solely on generic research ethics frameworks and highlighted the need for dedicated, context-sensitive ethics codes and guidance that can support ethically responsible research during times of crisis and fragility.

To promote the ethical conduct of essential research in challenging contexts, the European Commission funded a project team to develop two ethics codes for use during times of crisis. The first of these, the PREPARED Code – A Global Code of Conduct for Research during Pandemics (PREPARED 2025a), was designed to address research ethics and integrity issues that emerge or are intensified during pandemics (Chatfield and Singh 2025). The second, and focus of this book, is entitled the TRUST Code Supplement – A Global Code of Conduct for Research in Fragile Settings (PREPARED 2025b). This code has a broader scope, aiming to address ethics and integrity challenges that may arise across a wide range of fragile contexts.

Ensuring the relevance and robustness of an ethics code requires that it be firmly grounded in real-world challenges reflective of diverse research types and settings. As an initial step, this necessitates the identification and analysis of research ethics- and integrity-related challenges to clarify both what issues must be addressed and how this might be achieved. Given the heterogeneity of fragile settings, affecting different populations in different ways, this task required extensive groundwork.

To help bring to light the research ethics and integrity challenges in fragile settings three studies were conducted:

1. a focused scoping review of relevant academic literature published in English
2. a scoping review–informed evidence mapping of relevant academic literature published in German, replicating the search terms from this English language scoping review (see Chap. 4)
3. an online survey exploring the experiences and perspectives of individuals who have conducted research in fragile settings – an anonymous survey that collected responses from 37 researchers, identifying ethics- and integrity-related risks and challenges they had encountered or observed in the field. Participants had direct experience working in contexts such as conflict or post-conflict zones, regions recently affected by natural disasters, resource-constrained refugee camps, and high-crime environments including informal settlements (see Chap. 5).

This chapter presents the focused scoping review of relevant academic literature published in English, beginning with a description of the methods.

3.2 Methods

3.2.1 Method Selection

The focused scoping review was conducted in accordance with established methodological guidance for scoping studies. The methods followed the five-step framework proposed by Arksey and O'Malley (2005), which provides a structured and transparent approach to mapping evidence across diverse research fields. Scoping reviews are widely recognised for their iterative and flexible nature, allowing methodological decisions to be refined in response to feasibility considerations and the specific demands of the research context (Arksey and O'Malley 2005; Levac et al. 2010; Colquhoun et al. 2014). The approach taken embraced this accepted flexibility while prioritising the synthesis of a broad body of evidence (Munn et al. 2018).

Crucially, a multidatabase search is typically considered ideal for scoping reviews with review materials sourced across at least three databases. Given the time constraints of working to a very tight deadline, this review was limited to two databases, but these were selected strategically to ensure representation across relevant research domains. For this reason, the review is described as a *focused* scoping review. All other core scoping review methodological elements were maintained, including a broad research question, predefined search terms, a rigorous dual-screening process and transparent data charting (Arksey and O'Malley 2005; Peters et al. 2020). This approach balanced methodological rigour with pragmatic constraints, while remaining consistent with recognised scoping review practice. The five steps recommended by Arksey and O'Malley (2005) are as follows:

1. Identifying the research question
2. Identifying the relevant studies
3. Study selection
4. Charting the data
5. Collating, Summarising and Reporting the Results

The following sections describe how each of the five steps was implemented in this review.

3.2.2 Step 1: Identifying the Research Question

In line with the need to identify real-world challenges, the scoping review sought to address the following research question:

> What are the challenges for research ethics and research integrity when undertaking research in fragile settings?

And the objectives were:

1. Identify specific examples of research ethics issues that arise and/or are exacerbated when research is conducted in fragile settings.
2. Identify specific examples of research integrity issues that arise and/or are exacerbated when research is conducted in fragile settings.
3. Theme the identified examples according to the nature and type of research ethics or research integrity challenge that they represent.

3.2.3 Step 2: Identifying Relevant Studies

In its *States of Fragility* report, the Organisation for Economic Co-operation and Development (2022) describes fragility as a "combination of exposure to risk and insufficient coping capacities of the state, system and/or communities to manage, absorb or mitigate those risks". Phenomena such as conflict, forced displacement, violent extremism and famine are both causes and consequences of fragility. In some cases, fragility affects entire countries, while in others it is concentrated in specific pockets within otherwise stable states. Although most fragile contexts are found in sub-Saharan Africa, significant numbers also exist across the Middle East, North Africa, Asia, Latin America and the Caribbean. It was thus necessary for the chosen search strategy to reflect a wide range of types of settings, including the following:

1. conflict or post-conflict settings
2. settings that had recently experienced natural disasters such as earthquakes, tsunamis, hurricanes or floods
3. settings that were or had been devastated by climate change
4. refugee camps in lower-income regions
5. settings characterised by high violence and crime.

All five create significant risks for those who live and work in such settings, making them highly risky places, even for the conduct of research. Nine separate search terms were selected to capture the breadth of these types of fragile settings, as shown in Table 3.1.

These terms were combined with the terms "research ethics" and "research integrity" to perform nine searches across each database using the Boolean query:

```
(research ethics) OR (research integrity) AND (type of fragile
setting)
```

Searches were conducted using two large, well-established and multidisciplinary databases, PubMed and Web of Science.

PubMed was selected because it indexes a large body of biomedical, public health and global health research, where ethics and integrity challenges in fragile and conflict-affected settings are frequently discussed.

Table 3.1 Fragile setting search terms

Search term	To capture the research ethics or integrity risks occurring in
1. Conflict	Any type of conflict or post-conflict setting
2. Disaster	Any type of disaster setting
3. Earthquakes, tsunamis, hurricanes or floods	Any of these natural disaster settings
4. War	Recently experienced or ongoing war situations
5. Refugees	Refugee populations living in low- or lower-middle-income settings
6. High crime	Settings characterised by high rates of criminal activity
7. Climate change	Settings devastated by climate change
8. Fragile	Any type of fragile setting
9. Humanitarian	Any region undergoing humanitarian interventions

Web of Science was selected because it complements the use of PubMed by providing broad coverage across interdisciplinary journals, including social science, ethics, policy and development studies that engage with matters of research governance.

Together, these two databases capture much of the peer-reviewed, indexed literature relevant to research ethics and integrity in fragile settings.

3.2.4 Step 3: Study Selection

For each of the nine searches, all results were assessed for eligibility according to the inclusion and exclusion criteria detailed in Table 3.2. Two reviewers undertook the screening by reading abstracts and the full text where appropriate. Following assessment by the first reviewer, a quality control check was performed by the second reviewer across all hits to verify eligibility.

Following assessment, eligible publications from each of the searches across both databases were indexed on Excel spreadsheets, with one Excel page per search term (18 pages in total). The number of eligible publications per search term found via each database is shown in Table 3.3.

Across the two databases, out of total number of 753 hits, 106 publications were deemed eligible. However, there were a relatively large number of duplications, both across search terms and across the two databases. Once the results were organised on a single Excel sheet, all duplications ($n = 40$) were removed, leaving a total of 66 eligible publications (see Appendix) to be included in the scoping review. Flow charts

Table 3.2 Inclusion and exclusion criteria

Criteria	Inclusion	Exclusion
Language	English	Non-English
Type of publication	All published literature types	
Nature of content	First/second-hand experiences in fragile settings, research studies or clinical trials, case studies, workshops, expert meetings, systematic reviews, surveys, conference reports etc.	Theoretical, not based on real-world experiences
Research ethics and integrity content	Some focus on research ethics and/or research integrity issues	Limited to statement about research ethics approval
Disease outbreaks	None unless associated with another type of fragility	Epidemics/pandemics such as Ebola, HIV, COVID-19
Fragile	Fragility of the setting for the research	Fragility as a state of being: for instance, in the elderly
Conflict	Research conducted in conflict or post-conflict settings	Conflicts of interest not related to fragile settings
Refugee	Low-income settings	High-income settings
War*	Research conducted in war or post-war settings	Historical challenges, for instance, in Nazi research

*While date of publication was not a limiting factor, publications that described historical violations of research ethics and integrity, prior to the introduction of governance systems (like the requirement for research ethics approval) were excluded

Table 3.3 Relevant number of publications per search term

Search term	PubMed (eligible/total hits)	Web of science (eligible/total hits)
1. "Conflict"	14/270	9/41
2. "Disaster"	18/42	9/13
3. "Earthquakes" OR "Tsunamis" OR "Hurricanes" OR "Floods"	4/14	0/0
4. "War"	5/125	3/23
5. "Refugees"	6/50	5/8
6. "High crime"	5/107	0/0
7. "Climate change"	0/2	0/4
8. "Fragile"	3/14	3/3
9. "Humanitarian"	17/31	5/6
TOTAL	**72/655**	**34/ 98**

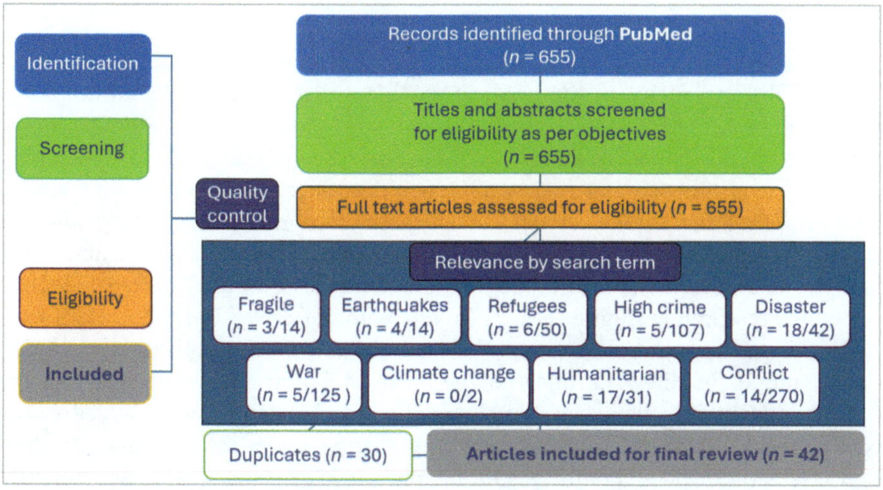

Fig. 3.1 Publications identified via PubMed

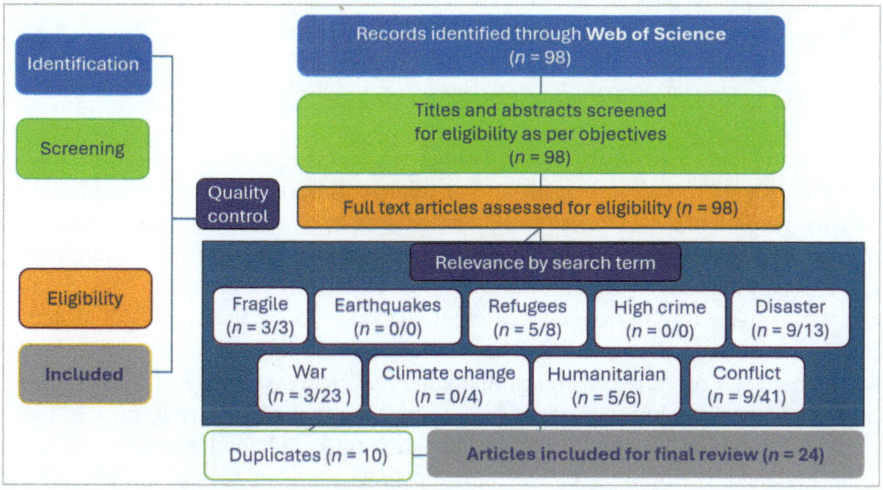

Fig. 3.2 Publications identified via web of science

representing the identification process of relevant literature are shown in Figs. 3.1 and 3.2.

3.2.5 Step 4: Charting the Data

Given the heterogeneity of the publications identified, which included empirical studies, reports, reviews and reflective accounts of experiences, the extraction of methods or an assessment of quality would not have been meaningful in this scoping review. Instead, the scoping review approach described by Arksey and O'Malley (2005) was modified for the very specific purpose of informing the development of a research ethics and integrity code. This required identification and extraction of the examples of research ethics and integrity issues that arise or are exacerbated when conducting research in fragile settings (in line with objectives 1 and 2).

For this task, each reviewer took the lead on half of the publications included while acting as quality controller for the other half. Full texts were read and the challenges identified for research ethics and research integrity were added to the Excel spreadsheet of 66 eligible review publications under the domains of "challenges for participants", "challenges for researchers", "research quality and integrity challenges", and "other challenges". A notes column was also included for any other information relevant to the data analysis or development of a code. Thus, the extracted data was sorted on the Excel spreadsheet under the following headings:

Reference	Type of setting	Challenges for participants	Challenges for researchers	Research quality and integrity challenges	Other challenges	Notes

Following data charting, the information extracted was examined through content analysis to identify and synthesise patterns related to the research ethics and research integrity challenges. Challenges related to similar concerns and concepts were grouped into descriptive categories, which were subsequently refined through analysis to generate broader themes. Analysis was conducted both within each domain (e.g., challenges for participants or challenges for researchers) and then across domains, as many challenges cut across multiple domains. For instance, power asymmetries between researchers and participants may increase the risk of coercion or undue influence for participants, while also reflecting wider structural inequities related to agenda-setting and the marginalisation of local knowledge.

To enhance credibility, theme development was conducted collaboratively between the two reviewers, with differences in interpretation resolved through discussion and re-examination of the original data. This led to the identification of the challenges for research ethics and research integrity described in the next section.

3.2.6 Step 5: Collating, Summarising and Reporting the Results

To inform the development of the ethics code and ensure that it addresses the real-world risks encountered in practice, it was necessary to identify as many challenges relating to research ethics and research integrity as possible. Thus, efforts to reveal challenges were extensive and a vast range were documented. Rather than listing every individual challenge identified in the English literature, this chapter synthesises the findings in line with the data analysis strategy. The identified risks have been themed and grouped into four categories: challenges for research participants, challenges for researchers, broader challenges for research ethics, and challenges for research quality and research integrity. What follows here is thus an overview of the principal challenges identified, rather than an exhaustive account of every issue reported in the literature.

3.3 Findings

3.3.1 Challenges for Research Participants

3.3.1.1 Compromised Informed Consent

Across the literature reviewed, challenges relating to informed consent were the most frequently cited. In fragile settings, consent processes may be compromised by a range of factors including trauma, crisis conditions, limited literacy, cultural and linguistic misunderstandings, time pressures, and confusion about the purpose and implications of research. In humanitarian and clinical contexts, participants may conflate research with aid or treatment, and therefore give their consent based on expectations of direct benefit. Additionally, power asymmetries between researchers and participants heighten the risk of coercion or undue influence. As a result, participants may agree to take part in research without fully understanding what participation entails or without genuine voluntariness.

Example Challenges

- decision-making capacity compromised due to factors like trauma, distress, injury or acute uncertainty
- urgency and uncertainty impacting the information provision process
- fear or mistrust of signing consent forms
- linguistic, literacy and cultural barriers
- confusion between research and aid/clinical care (therapeutic or humanitarian misconception)
- perceived obligation to participate in order to receive aid or healthcare
- fear of consequences of not participating (e.g., investigators accompanied by armed guards).

(Key references: Leaning 2001; NRC 2002; Van der Loos et al. 2014; Chiumento et al. 2016; Chiumento et al. 2017; Tansey et al. 2017; Perakslis 2018; Kohrt et al. 2019; Bruno and Haar 2020; Fox et al. 2020; Seagle et al. 2020; Cremin et al. 2021; Poole et al. 2021; Ataullahjan et al. 2022; Kurihara et al. 2022; Abeysinghe and Leppold 2023; Adams et al. 2024; Burke et al. 2024; Davidson et al. 2024; Ng et al. 2024; Thomson et al. 2024; Tiruneh et al. 2024; Baker et al. 2025; Vogel and Musamba 2025)

3.3.1.2 Psychological Harm and Retraumatisation

The risk of psychological harm is consistently identified as a major ethics challenge for participants, particularly when research involves the exploration of topics such as violence, loss or displacement. Interactions with the researchers can trigger emotional distress, revive traumatic memories, or contribute to emotional exhaustion. This risk is heightened when sensitive experiences are repeatedly elicited across multiple studies without adequate support mechanisms.

Example Challenges

- recalling traumatic experiences during interviews
- emotional exhaustion from repeated questioning
- lack of psychosocial support or referral pathways
- insufficient researcher training in trauma-informed methods.

(Key references: Schopper et al. 2009; Refsdal 2014; Contractor 2016; Mezinska et al. 2016; Tansey et al. 2017; Seagle et al. 2020; Alburo-Cañete 2021; Kostovicova and Knott 2022; Thomas et al. 2022; Adams et al. 2024; Thomson et al. 2024; Baker et al. 2025; Vogel 2025)

3.3.1.3 Exploitation and Unfair Benefit Distribution

While research in crisis contexts often generates benefits for researchers from high-income countries and their institutions, participants in the crisis zones may be exposed to social, emotional or physical risks without corresponding meaningful benefits. Moreover, participants may enrol in studies with expectations of benefits that are ultimately not met. Such unmet expectations, together with perceived inequities in how research benefits and burdens are distributed, contribute to experiences of exploitation and are recognised drivers of mistrust toward researchers among affected individuals and communities.

Example Challenges

- fleeting visits by foreign researchers and fleeting interactions with participants
- extractive research practices (e.g., stories are exploited with little reciprocity)
- participants receiving no feedback
- research agendas misaligned with participant/local community needs
- limited local benefit or capacity building.

(Key references: Leaning 2001; NRC 2002; Schopper et al. 2009; Stratton 2012; Van der Loos et al. 2014; Mezinska et al. 2016; O'Mathúna and Siriwardhana 2017; Schiltz and Büscher 2018; Bruno and Haar 2020; Seagle et al. 2020; Pincock and Bakunzi 2021; Kostovicova and Knott 2022; Ebubedike et al. 2023; Ng et al. 2024; Franc and O'Mathúna 2025)

3.3.1.4 Privacy and Confidentiality

Maintaining confidentiality can be particularly challenging in fragile settings where structural, political and resource factors can hamper the usual methods of ensuring privacy and data security. Further, it may be difficult to truly anonymise data where participants can be easily identified through community size, social roles or distinctive experiences. Breaches of privacy may have severe consequences, exposing participants to stigma, violence or political reprisal, linking privacy concerns directly to personal safety.

Specific Challenges

1. difficulty anonymising data in small or close-knit communities
2. insecure data storage or transmission
3. interviews conducted in non-private spaces
4. fear of identification by authorities or armed actors.

(Key references: Leaning 2001; NRC 2002; Tansey et al. 2017; Chiumento et al. 2018; Perakslis 2018; Seagle et al. 2020; Poole et al. 2021; Ataullahjan et al. 2022; Thomas et al. 2022; Thomson et al. 2024; Baker et al. 2025)

3.3.1.5 Physical and Security Risks Associated with Participation

Also mentioned frequently across the literature is the fact that, in some contexts, participation in research can expose individuals to physical or security risks. For instance, being seen with researchers or disclosing sensitive information may increase the likelihood of surveillance, targeting or retaliation, especially in conflict-affected or politically unstable settings. Standard institutional risk assessment processes often prioritise researcher safety, with insufficient attention to participant-specific risks.

Example Challenges

- risk of targeting, retaliation or surveillance
- visiting unsafe research sites or travelling across unsafe regions for research participation
- disclosure of politically or socially sensitive information
- inadequate risk assessment for participant safety.

(Key references: Schopper D 2009; Bennouna et al. 2017; Tansey et al. 2017; Seagle et al. 2020; Almeida 2021; Kurihara et al. 2022; Thomas et al. 2022; Thomson et al. 2024; Vogel 2025)

3.3.1.6 Other Challenges for Participants

Community-Level Harms

A number of publications in the reviewed literature highlight that participation in research can produce unintended social consequences for participants. Being identified as a research participant may reinforce existing marginalisation, or generate tension, jealousy or suspicion within families and communities (Mezinska et al. 2016; Bennouna et al. 2017; Chiumento et al. 2017; Seagle et al. 2020; Almeida 2021; Thomas et al. 2022).

Paternalistic Exclusion

While many studies describe participants in fragile settings as vulnerable due to their exposure to conflict, displacement, poverty, trauma and so on, other authors caution against the paternalistic categorisation of entire populations as vulnerable. The routine categorisation of whole groups as vulnerable may lead to unfair exclusion of individuals who are capable of consenting (Ng et al. 2024) and who actively want to contribute to the research (Kostovicova and Knott 2022).

Research Fatigue

Finally, several publications draw attention to the cumulative burden experienced by participants in heavily researched settings. Populations affected by crises, displacement or conflicts are often approached repeatedly by multiple research teams and asked similar questions, frequently about distressing or traumatic experiences. This can result in fatigue, frustration and emotional strain (Refsdal 2014; Mezinska et al. 2016; Schiltz and Büscher 2018; Bentele and Herzog 2023).

3.3.2 Challenges for Researchers

3.3.2.1 Physical Safety and Security Risks

Safety and security concerns are among the most frequently identified ethical challenges for researchers in the reviewed literature. Many publications document physical security risks, including exposure to violence, intimidation or surveillance, particularly in conflict-affected or politically sensitive contexts. Even well-prepared research teams may encounter unforeseen threats, as manifestations of disaster and crisis are often highly variable in settings subject to geological, medical, political or military instability. The literature further indicates that security risks may be gendered and that prolonged engagement in such environments can foster a normalisation of risk within research cultures, potentially obscuring the need for caution.

Example Challenges

- travelling or working in difficult or dangerous and unpredictable settings
- community mistrust and threats from armed actors
- suspicion from authorities about who the researchers "really are" and why they are navigating a conflict zone
- researchers facing physical violence
- female scholars facing sexual harassment or threats during fieldwork.

(Key references: Helbardt et al. 2010; O'Mathúna 2010; O'Mathúna 2015; Chiumento et al. 2016; Chiumento et al. 2017; Tansey et al. 2017; Chiumento et al. 2018; Schiltz and Büscher 2018; Campbell et al. 2019; Bruno and Haar 2020; Cremin et al. 2021; Krause 2021; Thomas et al. 2022; Vogel 2025; Vogel and Musamba 2025)

3.3.2.2 Emotional Challenges and Moral Distress

Beyond physical risk, researchers in crisis settings often experience vicarious trauma along with emotional exhaustion, distress and psychological strain. This can arise from sustained exposure to suffering, violence, loss and injustice, particularly when research involves sensitive topics such as displacement, trauma or gender-based violence. Many researchers also experience a form of moral distress when they encounter urgent needs, but institutional protocols, research roles or contextual constraints prevent them from responding to participants' needs, intervening in harmful situations or addressing injustices that they witness. Further, inexperienced researchers may be ill-prepared, or those who work for prolonged periods in these environments may become gradually habituated and thus fail to protect themselves adequately from the emotional fallout.

Example Challenges

- witnessing and reporting on the human suffering and physical destruction of a disaster
- repeated deep listening to narratives about atrocities, loss and pain, and observations of its daily impacts
- frequently receiving requests for help from participants who are in dire need
- psychological injury from inadequate protection and self-care.

(Key references: Helbardt et al. 2010; O'Mathúna 2010; Vervliet et al. 2015; Chiumento et al. 2016; Mezinska et al. 2016; Chiumento et al. 2017; Eriksen 2017; Campbell et al. 2019; Cremin et al. 2021; Baker et al. 2025; Vogel and Musamba 2025)

3.3.2.3 Researcher Preparedness

The literature identifies insufficient researcher preparedness for work in fragile settings as a significant challenge. Researchers are often described as inadequately trained for work in unstable environments, while time pressures can limit opportunities for meaningful familiarisation with local conditions. Researchers may also be unfamiliar with local cultural norms, power dynamics and community contexts, and while they will likely work with traumatised participants, they may lack the skills for trauma-informed interactions. Further, crisis research can require researchers to adapt methods rapidly and to take ethically complex decisions that require more than their existing expertise. The literature further highlights capacity gaps within research teams, including a lack of adequately trained local researchers. Thus, the need for appropriately skilled team members, ongoing training and support are frequently emphasised.

Example Challenges

- insufficient training for dealing with traumatised participants
- need for real-time ethical decision-making
- uncertainty about appropriate ethical action in unstable settings
- lack of local researchers with relevant skills for fieldwork.

(Key references: O'Mathúna 2015; Hunt et al. 2016; O'Mathúna and Siriwardhana 2017; Tansey et al. 2017; Kohrt et al. 2019; Bruno and Haar 2020; Alburo-Cañete 2021; Thomas et al. 2022; Abeysinghe and Leppold 2023; Bentele and Herzog 2023; Adams et al. 2024; Burke et al. 2024; Abeysinghe et al. 2025)

3.3.2.4 Navigating the Ethics Approval Process

Another recurring challenge for researchers appears to lie in the perceived mismatch between institutional ethics review processes and the realities of research in fragile settings (see also Chap. 2). Procedural ethics approval processes often fail to capture the complexities of research in fragile settings, where researchers are frequently required to make context-specific ethical decisions on the spot. Many authors report that formal ethics procedures can be rigid and bureaucratic, with research ethics committee (REC) members who have insufficient understanding of research in crisis or insecurity settings. The literature also describes difficulties obtaining timely approval for new proposals or for modifying protocols in response to rapidly changing circumstances.

Example Challenges

- lengthy approval processes threatening project feasibility
- REC members' narrow focus, which can appear overly risk- and litigation-averse

- rigid REC requirements limiting the researcher's ability to engage with situated ethics.

(Key references: O'Mathúna and Siriwardhana 2017; Aung et al. 2019; Habib 2019; Fox et al. 2020; Cremin et al. 2021; Krause 2021; Ataullahjan et al. 2022; Adams et al. 2024; Abeysinghe et al. 2025; Baker et al. 2025)

3.3.3 Broader Challenges for Research Ethics

3.3.3.1 Fragile or Disrupted Ethics Governance Systems

Several authors report the absence, limited capacity or dysfunction of local and national RECs, including situations in which review boards are nonexistent, suspended or unable to operate effectively due to conflict or disaster-related disruption (O'Mathúna 2015; Chiumento et al. 2016; Sombié et al. 2017). Insecurity, damaged infrastructure and shortages of trained personnel further constrain the functioning of ethics review systems and limit the feasibility of certain types of research, including medical and humanitarian research (Tiruneh et al. 2024). Weak or inconsistent ethical oversight is also evidenced through low levels of reporting on ethics approval processes, ethics waivers and informed consent in published research (Hussain and Elmusharaf 2019; Ataullahjan et al. 2022). These gaps have been attributed to a combination of lenient local standards, heightened vulnerability of affected populations, and the collapse of institutional infrastructure following crises (Van der Loos et al. 2014). In volatile and rapidly shifting conflict environments, researchers may also find it difficult or impossible to obtain permissions from authorities with sufficient legitimacy or stability, further hindering effective governance (Vogel and Musamba 2025).

3.3.3.2 Power Asymmetries in Research Partnerships

Power asymmetries in research relationships are identified as a challenge by several authors, particularly within North-South research partnerships. For instance, there can be unequal control over funding, infrastructure and agenda-setting, with resources often flowing through elite institutions (Bentele and Herzog 2023). Power imbalances can raise profound ethical questions about conducting research among populations under duress, where participation may occur in contexts of limited choice, with heightened vulnerability and risk (Helbardt et al. 2010). Power inequalities are also experienced at the interpersonal and community level, where stakeholders may become dependent on decisions made by researchers or institutions with greater authority and mobility (Ng et al. 2024). Structural hierarchies within research collaborations further reinforce these dynamics, with researchers from the Global North commonly taking the lead in the research (Thomas et al. 2022). These imbalances are compounded by epistemic inequalities, which often see the validity of Global South knowledge systems marginalised or negated by dominant Global North epistemologies (Cremin et al. 2021). Collectively, these issues foreground persistent concerns around access, trust, positionality and power in the conduct of research in fragile settings (Vogel 2025).

3.3.4 Challenges for Research Quality and Research Integrity

3.3.4.1 Compromised Methodological Rigour

Research in crisis settings is often shaped by insecurity, urgency, displacement, weak infrastructures and access limitations that compromise methodological rigour across all stages of the research process. These practical, methodological and ethical constraints can make it difficult to generate valid and reliable data, particularly given the challenges in accessing unbiased information, the near impossibility of achieving representative sampling, and the risk of contaminated or distorted evidence.

Example Challenges

1. inconsistent and unsystematic collection and analysis
2. methods and instruments with limited validity when used in oral cultures
3. data collection dependent upon whom one can get close to
4. recruitment dependent upon the accessibility, cost and malleability of potential participants.

(Key references: Helbardt et al. 2010; Chiumento et al. 2016; Mezinska et al. 2016; Sen et al. 2016; Bennouna et al. 2017; Chiumento et al. 2017; Sombie et al 2017; Campbell et al. 2019; Kohrt et al. 2019; Xi et al. 2019; Seagle et al. 2020; Abeysinghe and Leppold 2023; Adams et al. 2024; Tiruneh et al. 2024)

3.3.4.2 Duplication and Lack of Coordination of Research

Across the literature, many authors note challenges associated with unnecessary duplication and poor coordination of research, with similar or overlapping research being repeatedly conducted in the same contexts, often in ignorance of existing or ongoing work (Chiumento et al. 2017; NRC 2002; Sombié et al. 2017; Ng et al. 2024). Weak coordination between research teams, institutions and funders furthers fragmented agendas and parallel data collection efforts, rather than coherent programmes of research (NRC 2002; Ng et al. 2024). Duplication and poorly coordinated efforts place additional burdens on already overstretched communities and systems, while failing to substantially strengthen the evidence base, thereby raising concerns about the responsible use of resources and the integrity of the research.

3.3.4.3 Other Research Integrity Matters

Aside from issues of research quality, relatively few challenges in the dataset can be viewed explicitly as matters of research integrity rather than research ethics. Those that are identified centre on bias in reporting, bias in stance and conflicts of interest. For instance, academic outputs may privilege particular perspectives, with publications exhibiting a bias towards one side of a conflict (Helbardt et al. 2010). Conflicts of interest are apparent where researchers design or frame studies in ways that support advocacy positions or fundraising objectives (Hunt et al. 2016). Relatedly, sanitised research accounts have been criticised for presenting unrealistic portrayals of orderly and unproblematic research processes, obscuring the ethical and methodological complexities of work in crisis settings (Chiumento et al. 2016). More broadly, the literature points to prejudice and polarisation as contaminating both scientific practice and ethical reasoning, making fair and balanced representation extremely difficult, especially in contexts of active or recent conflict (Sen et al. 2016).

3.4 Discussion

This focused scoping review sought to address the research question: *What are the challenges for research ethics and research integrity when undertaking research in fragile settings?* To this end, it aimed to identify real-world examples of ethics- and integrity-related challenges, and to theme the identified examples in a manner that reflects the nature and scope of the issues encountered. The findings demonstrate that research in fragile settings is characterised by a complex interplay of ethical risks, methodological constraints and structural weaknesses that affect research participants, researchers, governance systems and the integrity of the research itself. These settings can be highly charged politicised environments, they can be geographically dangerous, and they might be under military rule. Researchers may face perilous situations within societies that are deeply divided, and with participants who have witnessed or experienced violence or have been deeply traumatised by preceding events. The scoping review fundamentally highlighted the need for ethical vigilance that is sensitive to the unique disarray that researchers encounter in fragile settings.

Consistent with the first objective (to identify specific examples of research ethics issues that arise and/or are exacerbated when research is conducted in fragile settings), the review revealed a wide range of such challenges. Many of these challenges, such as compromised informed consent, risks of psychological harm, threats to privacy and participant exploitation, are not unique to fragile contexts. However, the literature highlights how fragility can magnify their severity, likelihood and consequences.

Informed consent emerged as the most consistently reported ethical challenge for participants. Across diverse fragile contexts, factors such as trauma, low literacy, linguistic barriers, urgency, fear and humanitarian dependency complicate participants' ability to provide fully informed and voluntary consent. The conflation of research with aid or clinical care further exacerbates this challenge, raising concerns about therapeutic or humanitarian misconception and the risk of undue influence. Risks of psychological harm, retraumatisation and community-level impacts were also prominent. These findings emphasise that the risk of harm from research in fragile settings extends beyond the immediate physical concerns and may include emotional exhaustion, social stigma and the longer-term erosion of trust.

The review also identified significant ethical challenges affecting researchers themselves. Physical insecurity, emotional distress, vicarious trauma and moral distress are recurring themes across the literature. While institutional ethics processes typically emphasise participant protection, the findings indicate that researcher well-being also needs to be supported. The emotional and moral dimensions of research in fragile settings are particularly salient. Researchers often operate in contexts of acute human need, where rigid role boundaries and logistical constraints prevent them from responding to the suffering that they witness. The review also highlights how inadequate preparation, limited training in trauma-informed methods and lack of contextual familiarity compound these challenges.

Beyond individual-level challenges, the review identifies systemic weaknesses in research ethics governance as a defining feature of fragile settings. The absence, disruption or limited capacity of local RECs can constrain meaningful oversight, while externally imposed review processes may lack contextual sensitivity or operational flexibility. Power asymmetries, particularly within North-South research partnerships, further complicate matters and raise persistent concerns about extractive research and epistemic injustice.

In relation to the second objective (to identify specific examples of research integrity issues that arise and/or are exacerbated when research is conducted in fragile settings), the review demonstrates that challenges to research integrity in fragile settings are closely intertwined with ethical and practical constraints. For the initial sorting of extracted challenges, factors that impact research quality were placed in the same domain as the challenges for research integrity. That is because matters of quality such as "rigorous methods" and "reliable findings" are central to the integrity of research (World Conferences on Research Integrity 2010). But it should be recognised that scientific quality is also an ethical requirement; research that lacks scientific validity is unethical because it wastes resources and exploits participants (CIOMS 2016: Guideline 1).

Upholding the quality of research in fragile settings is a major challenge for researchers. Fragile and humanitarian contexts are inherently unpredictable and therefore imbue the research process with uncertainty. Researchers may experience limited access caused by restrictions on movement, checkpoints and other security measures that can hinder research activities. Data collection in unstable environments can be challenging, with barriers to data collection including ongoing displacement, armed conflict, instability and lack of security. This makes it dangerous and difficult for researchers to access certain areas or populations, and results in incomplete or compromised data. The urgency of the research in response to emergencies means that implementation cycles are often short and decisions about adaptions need to be made quickly.

Finally, while the preceding analysis may read like a catalogue of ethical shortcomings, it is important to emphasise that this review deliberately focused on identifying *challenges* to research ethics and research integrity, rather than good practice. Therefore, it does not reflect the full breadth or quality of research conducted in fragile settings. Many researchers operate in complex and often dangerous environments, frequently at significant personal risk, and are motivated by a genuine commitment to contribute positively. However, good intentions alone are insufficient to ensure ethical research practice. Ethical failures more often arise from inadequate preparation, limited training, insufficient institutional support and broader structural constraints, rather than from deliberate wrongdoing. Recognising these factors is essential for developing guidance and systems that better support researchers in conducting ethically robust research in fragile and crisis-affected contexts.

3.5 Study Limitations

The principal limitation of this review is related to the sources of the literature included, which were drawn from only two databases. As a result, some relevant publications may not have been captured, particularly work not fully indexed in PubMed or Web of Science. This includes grey literature and research from disciplines such as anthropology, which are often published in outlets beyond these databases. Consequently, the review reflects key themes within the indexed academic literature rather than the full range of perspectives on research ethics and integrity in fragile settings. In addition, much of the literature centres on the views of researchers, with comparatively limited representation of participant or local community perspectives, which remain underexplored in the available evidence.

3.6 Conclusion

To provide a robust empirical foundation for a new ethics code for research in fragile settings, this scoping review examined the challenges for research ethics and research integrity when undertaking research in fragile settings. The findings revealed that fragility can intensify ethical risks profoundly, affecting research participants, researchers, governance systems and the integrity of the research itself. Further, conditions of instability, uncertainty, insecurity, trauma and inequality complicate the application of standard ethics governance measures.

Across the literature, ethical challenges for participants such as compromised informed consent, heightened risks of harm and threats to confidentiality are persistent and interrelated. At the same time, researchers can face significant physical, emotional and moral burdens, while weak or disrupted ethics governance and power asymmetries within research partnerships further constrain ethical practice. Challenges to research integrity largely reflect contextual and structural constraints rather than individual misconduct, underscoring the close relationship between ethics, methodology and research quality in fragile settings.

Overall, these challenges have illuminated the fundamental need for guidance to ensure responsible practice in fragile settings. Research ethics and integrity measures should be responsive and commensurate with the requisites occasioned by the fragility of the settings and support the safety and wellbeing of participants, communities and researchers. In environments already ravaged by instability, trauma and complexity, the urgent need for effective ethical oversight of research becomes even more pertinent. Hence the need to develop the Trust Code Supplement to guide ethically responsible, high-quality research in fragile and crisis-affected environments.

Appendix: Literature Included in the Focused Scoping Review

Key

D = Disaster; F = Fragile; C = Conflict; H = Humanitarian; ETHF = Earthquake, tsunami, hurricane or flood; W = War; R = Refugees; HC = High crime

	Reference	Type
1	Abeysinghe S, Leppold C (2023) Ethical research practice in health and disasters. International Journal of Disaster Risk Reduction 92:103728. https://doi.org/10.1016/j.ijdrr.2023.103728	D
2	Abeysinghe S, Honda K, Ozaki A, et al (2025) Disciplinary research ethics cultures in disaster settings: understandings from researchers after 3.11 in Japan. Journal of Empirical Research on Human Research Ethics 21(1–2):16–24. https://doi.org/10.1177/15562646251399397	D
3	Abu-Rmeileh N (2019) Challenges facing research ethics in fragile settings. Indian Journal of Medical Ethics 4(4)336–337. https://doi.org/10.20529/IJME.2019.067	F
4	Adams RM, Evans CM, Peek L (2024) Defining, collecting, and sharing perishable disaster data. Disasters 48: e12592. https://doi.org/10.1111/disa.12592	D
5	Alburo-Cañete KZ (2021) PhotoKwento: co-constructing women's narratives of disaster recovery. Disasters 45:887–912. https://doi.org/10.1111/disa.12448	D, ETHF
6	Almeida MC e, Harris A (2021) The conflict sensitivity principle: can best practice in conflict research fill the ethics gap in terrorism and counterterrorism research practice? Terrorism and Political Violence 33(2):381–396. https://doi.org/10.1080/09546553.2021.1880159	C
7	Ataullahjan A, Lo S, Haaris Aziz M et al. (2022) Why we need stricter oversight of research involving human subjects affected by conflict. AMA Journal of Ethics 24(6):E518–529. https://doi.org/10.1001/amajethics.2022.518	H, C

(continued)

(continued)

	Reference	Type
8	Aung MN, Murray V, Kayano R (2019) Research methods and ethics in health emergency and disaster risk management: the result of the Kobe Expert Meeting. International Journal of Environmental Research and Public Health 16(5):770. https://doi.org/10.3390/ijerph16050770	D
8	Baker S, Burke R, Cabiles B et al (2025) Navigating institutional ethics processes: insights from higher degree by research students and supervisors doing research in fragile contexts. Qualitative Research 25(5):963–983. https://doi.org/10.1177/14687941241288178	F
10	Barber R (2019) The ethics of research in humanitarian action. In: Harris V (ed) Ethics in a crowded world: globalisation, human movement and professional ethics. Emerald Publishing Limited, pp 69–85. https://doi.org/10.1108/S1529-209620190000022006	H
11	Bennouna C, Mansourian H, Stark L (2017) Ethical considerations for children's participation in data collection activities during humanitarian emergencies: a Delphi review. Conflict and Health 11(1):5. https://doi.org/10.1186/s13031-017-0108-y	H
12	Bentele U, Herzog M (2023) Conflict-sensitive research: bringing research ethics into practice. Development in Practice 33(5):656–662. https://doi.org/10.1080/09614524.2023.2213860	C
13	Bloswick A, Mustra D, Harasymiv O, Dubov A (2022) Facilitating Ukrainian refugees' continued participation in clinical trials. Hastings Center Report 52:6–8. https://doi.org/10.1002/hast.1390	W, C, R
14	Bruno W, Haar RJ (2020) A systematic literature review of the ethics of conducting research in the humanitarian setting. Conflict and Health 14(1):27. https://doi.org/10.1186/s13031-020-00282-0	H
15	Burke R, Baker S, Molla T et al (2024) How do higher degree research students and supervisors navigate ethics-in-practice for educational research in sensitive or 'fragile'contexts? British Educational Research Journal 50(2):837–854. https://doi.org/10.1002/berj.3945	F
16	Calain P (2018) The Ebola clinical trials: a precedent for research ethics in disasters. Journal of Medical Ethics 44(1):3–8. https://doi.org/10.1136/medethics-2016-103474	D
17	Campbell R, Goodman-Williams R, Javorka M (2019) A trauma-informed approach to sexual violence research ethics and open science. Journal of Interpersonal Violence 34(23–24):4765–4793. https://doi.org/10.1177/0886260519871530	HC
18	Chan EYY, Wright K, Parker M (2019) Health-emergency disaster risk management and research ethics. Lancet 12;393(10167):112–113. https://doi.org/10.1016/s0140-6736(18)33126-x	D
19	Chiumento A, Khan MN, Rahman A, Frith L (2016) Ethical challenges to research in post-conflict settings. Developing World Bioethics 16:15–28. https://doi.org/10.1111/dewb.12076	H, C
20	Chiumento A, Machin L, Rahman A, Frith L (2018) Online interviewing with interpreters in humanitarian contexts. International Journal of Qualitative Studies on Health and Well-Being 13(1). https://doi.org/10.1080/17482631.2018.1444887	H, W, C

(continued)

(continued)

	Reference	Type
21	Chiumento A, Rahman A, Frith L et al. (2017) Ethical standards for mental health and psychosocial support research in emergencies: review of literature and current debates. Global Health 13(8). https://doi.org/10.1186/s12992-017-0231-y	D, C
22	Contractor Q (2016) Fieldwork and social science research ethics. Indian Journal of Medical Ethics 1(5):22–23. https://doi.org/10.20529/IJME.2008.008	H, C
23	Cremin H, Aryoubi H, Hajir B et al (2021) Post-abyssal ethics in education research in settings of conflict and crisis: stories from the field. British Educational Research Journal 47(4):1102–1119. https://doi.org/10.1002/berj.3712	C
24	Davidson N, Hammarberg K, Fisher J (2024) Ethical considerations in research with people from refugee and asylum seeker backgrounds: a systematic review of national and international ethics guidelines. Journal of Bioethical Inquiry 21(2):261–284. https://doi.org/10.1007/s11673-023-10297-w	R
25	Ebubedike M, Akanji T, Kunock AI, Fox A (2023) Ethics for educational research in regions of protracted armed conflict and crisis: a participatory community project in the Lake Chad region. Community Development Journal 58(1):102–120. https://doi.org/10.1093/cdj/bsac040	C
26	Eckenwiler L, Pringle J, Boulanger R, Hunt M (2015) Real-time responsiveness for ethics oversight during disaster research. Bioethics 29(9):653–661. https://doi.org/10.1111/bioe.12193	D, ETHF
27	Eriksen C (2017) Research ethics, trauma and self-care: reflections on disaster geographies. Australian Geographer 48(2):273–278. https://doi.org/10.1080/00049182.2016.1230001	D
28	Ferreira RJ, Buttell F, Cannon C (2018) Ethical issues in conducting research with children and families affected by disasters. Current Psychiatry Reports 20(6):42. https://doi.org/10.1007/s11920-018-0902-2	D
29	Fox A, Baker S, Charitonos K et al (2020) Ethics-in-practice in fragile contexts: research in education for displaced persons, refugees and asylum seekers. British Educational Research Journal 46(4):829–847. https://doi.org/10.1002/berj.3618	F
30	Franc JM, O'Mathúna D (2025) Navigating research ethics for prehospital and disaster medicine. Prehospital and Disaster Medicine 40(3):125–128. https://doi.org/10.1017/S1049023X25101234	D
31	Habib RR (2019) Ethical, methodological, and contextual challenges in research in conflict settings: the case of Syrian refugee children in Lebanon. Conflict and Health 13:29. https://doi.org/10.1186/s13031-019-0215-z	HC, C, R
32	Helbardt S, Hellmann-Rajanayagam D, Korff R (2010) War's dark glamour: ethics of research in war and conflict zones. Cambridge Review of International Affairs 23(2):349–369. https://doi.org/10.1080/09557571003752688	W, C
33	Hunt M, Tansey CM, Anderson J at al (2016) The challenge of timely, responsive and rigorous ethics review of disaster research: views of research ethics committee members. PLoS ONE 11(6):e0157142 https://doi.org/10.1371/journal.pone.0157142	D, ETHF
34	Hunt MR, Anderson JA, Boulanger RF (2012) Ethical implications of diversity in disaster research. American Journal of disaster medicine 7(3):211–221. https://doi.org/10.5055/ajdm.2012.0096	D

(continued)

(continued)

	Reference	Type
35	Hussein G, Elmusharaf K (2019) Mention of ethical review and informed consent in the reports of research undertaken during the armed conflict in Darfur (2004–2012): a systematic review. BMC Medical Ethics 20:40. https://doi.org/10.1186/s12910-019-0377-7	H, W, C
36	Kohrt BA, Mistry AS, Anand N et al (2019) Health research in humanitarian crises: an urgent global imperative. BMJ Global Health 4(6):e001870. https://doi.org/10.1136/bmjgh-2019-001870	H, W, C
37	Kostovicova D, Knott E (2022) Harm, change and unpredictability: the ethics of interviews in conflict research. Qualitative Research 22(1):56–73. https://doi.org/10.1177/1468794120975657	C
38	Krause J (2021) The ethics of ethnographic methods in conflict zones. Journal of Peace Research 58(3):329–341. https://doi.org/10.1177/0022343320971021	HC, C
39	Kurihara C, Crawley FP, Baroutsou V et al (2022) The continuation of clinical trials in times of war: a need to develop ethics and situationally adaptive clinical research guidelines. Frontiers in Medicine 9:966220. https://doi.org/10.3389/fmed.2022.966220	H, W
40	Leaning J (2001) Ethics of research in refugee populations. The Lancet 357(9266):1432–1433	R
41	Mezinska S, Kakuk P, Mijaljica G et al (2016) Research in disaster settings: a systematic qualitative review of ethical guidelines. BMC Medical Ethics 17:62. https://doi.org/10.1186/s12910-016-0148-7	D
42	National Research Council (2002) Research ethics in complex humanitarian emergencies: summary of a workshop. National Academies Press (US), Washington DC. https://doi.org/10.17226/10481	C, H, R
43	Ng SH, Kaur S, Cheah PY et al (2024) Migration health ethics in Southeast Asia: a scoping review. Wellcome Open Research 4(8):391. https://doi.org/10.12688/wellcomeopenres.19572.3	HC, R
44	O'Mathúna DP (2010) Conducting research in the aftermath of disasters: ethical considerations. Journal of Evidence-Based Medicine 3:65–75. https://doi.org/10.1111/j.1756-5391.2010.01076.x	D
45	O'Mathúna D (2015) Research ethics in the context of humanitarian emergencies. Journal of Evidence-Based Medicine 8:31–35. https://doi.org/10.1111/jebm.12136	H, D
46	O'Mathúna D, Siriwardhana C (2017) Research ethics and evidence for humanitarian health. Lancet 18;390(10109):2228–2229. https://doi.org/10.1016/s0140-6736(17)31276-x	H
47	Perakslis ED (2018) Using digital health to enable ethical health research in conflict and other humanitarian settings. Conflict and Health 12(1):23. https://doi.org/10.1186/s13031-018-0163-z	H, C
48	Pincock K, Bakunzi W (2021) Power, participation, and 'peer researchers': addressing gaps in refugee research ethics guidance. Journal of Refugee Studies 34(2):2333–2348. https://doi.org/10.1093/jrs/feaa060	R

(continued)

(continued)

	Reference	Type
49	Poole DN, Raymond NA, Berens J et al (2021) A combination sampling approach for epidemiologic research in humanitarian settings: a case analysis of a study of depressive disorder prevalence among refugees in Greece. BMC Public Health 21:272. https://doi.org/10.1186/s12889-021-10342-6	H, R
50	Refsdal NO (2014) Experiences from coordinating research after the 2011 terrorist attacks in Norway. European Journal of Psychotraumatology 5(1). https://doi.org/10.3402/ejpt.v5.23215	D
51	Schiltz J, Büscher K (2018) Brokering research with war-affected people: the tense relationship between opportunities and ethics. Ethnography 19(1):124–146. https://doi.org/10.1177/1466138116664542	W
52	Schopper D, Dawson A, Upshur R et al (2015) Innovations in research ethics governance in humanitarian settings. BMC Medical Ethics 26(16):10. https://doi.org/10.1186/s12910-015-0002-3	F, H
53	Schopper D, Upshur R, Matthys F et al (2009) Research ethics review in humanitarian contexts: the experience of the independent ethics review board of Médecins Sans Frontières. PLOS Medicine 6(7):e1000115. https://doi.org/10.1371/journal.pmed.1000115	H
54	Seagle EE, Dam AJ, Shah PP et al (2020) Research ethics and refugee health: a review of reported considerations and applications in published refugee health literature, 2015–2018. Conflict and Health 14(1):39. https://doi.org/10.1186/s13031-020-00283-z	R
55	Sen K, Hussain H, Faisal WA (2016) Ethics in times of conflict: some reflections on Syria, in the backdrop of Iraq: BMJ Global Health. 1:e000149. https://gh.bmj.com/content/1/3/e000149	D, C
56	Sombié I, Aidam J, Montorzi G (2017) Evaluation of regional project to strengthen national health research systems in four countries in West Africa: lessons learned. Health Research Policy and Systems 15(Suppl 1):46. https://doi.org/10.1186/s12961-017-0214-8	F, C
57	Stratton S J (2012) Ethics in EMS and disaster research. Prehospital and Disaster Medicine 27(06):495. https://doi.org/10.1017/S1049023X12001495	D
58	Tansey CM, Anderson J, Boulanger RF et al (2017) Familiar ethical issues amplified: how members of research ethics committees describe ethical distinctions between disaster and non-disaster research. BMC Medical Ethics 18(1):44. https://doi.org/10.1186/s12910-017-0203-z	H, D, ETHF
59	Thomas SN, Weber S, Bradbury-Jones C (2022) Using participatory and creative methods to research gender-based violence in the Global South and with Indigenous communities: findings from a scoping review. Trauma, Violence, & Abuse 23(2):342–355. https://doi.org/10.1177/1524838020925775	HC
60	Thomson B, Mehta S, Robinson C (2024) Scoping review and thematic analysis of informed consent in humanitarian emergencies. BMC Medical Ethics 25:135. https://doi.org/10.1186/s12910-024-01125-w	H
61	Tiruneh G, Yilma M, Wakuma B et al (2024) Compliance with research ethics in epidemiological studies targeted to conflict-affected areas in Western Ethiopia: validity of informed consent (VIC) by information comprehension and voluntariness (ICV). BMC Medical Ethics 25(1):9. https://doi.org/10.1186/s12910-024-01003-5	C

(continued)

(continued)

	Reference	Type
62	Van der Loos KI, Longstaff H, Virani A, Illes J (2014) Consent in escrow. Journal of Law and the Biosciences 2(1):69–78. https://doi.org/10.1093/jlb/lsu031	D
63	Vervliet M, Rousseau C, Broekaert E, Derluyn I (2015) Multilayered ethics in research involving unaccompanied refugee minors. Journal of Refugee Studies 28(4):468–485. https://doi.org/10.1093/jrs/feu039	R
64	Vogel B (2025) Mind the ethics gap: embedding research ethics into student fieldtrips to conflict and development settings. Politics 45(4):655–664. https://doi.org/10.1177/02633957251344550	C
65	Vogel C, Musamba J (2025) Towards a politics of collaborative worldmaking: ethics, epistemologies and mutual positionalities in conflict research. Ethnography 26(3):603–623. https://doi.org/10.1177/14661381221090895	C
66	Xi Y, Chen R, Gillespie AL et al (2019) Mental health workers perceptions of disaster response in China. BMC Public Health 19(1):11. https://doi.org/10.1186/s12889-018-6313-9	D

References

Abeysinghe S, Leppold C (2023) Ethical research practice in health and disasters. Int J Disaster Risk Reduct 92:103728. https://doi.org/10.1016/j.ijdrr.2023.103728

Abeysinghe S, Honda K, Ozaki A et al (2025) Disciplinary research ethics cultures in disaster settings: understandings from researchers after 3.11 in Japan. J Empir Res Hum Res Ethics 21(1–2):16–24. https://doi.org/10.1177/15562646251399397

Adams RM, Evans CM, Peek L (2024) Defining, collecting, and sharing perishable disaster data. Disasters 48:e12592. https://doi.org/10.1111/disa.12592

Alburo-Cañete KZ (2021) PhotoKwento: co-constructing women's narratives of disaster recovery. Disasters 45:887–912. https://doi.org/10.1111/disa.12448

Almeida MC, Harris A (2021) The conflict sensitivity principle: can best practice in conflict research fill the ethics gap in terrorism and counterterrorism research practice? Terror Polit Violence 33(2):381–396. https://doi.org/10.1080/09546553.2021.1880159

Arksey H, O'Malley L (2005) Scoping studies: towards a methodological framework. Int J Soc Res Methodol 8(1):19–32. https://doi.org/10.1080/1364557032000119616

Ataullahjan A, Lo S, Haaris Aziz M et al (2022) Why we need stricter oversight of research involving human subjects affected by conflict. AMA J Ethics 24(6):E518–E529. https://doi.org/10.1001/amajethics.2022.518

Aung MN, Murray V, Kayano R (2019) Research methods and ethics in health emergency and disaster risk management: the result of the Kobe Expert Meeting. Int J Environ Res Public Health 16(5):770. https://doi.org/10.3390/ijerph16050770

Baker S, Burke R, Cabiles B et al (2025) Navigating institutional ethics processes: insights from higher degree by research students and supervisors doing research in fragile contexts. Qual Res 25(5):963–983. https://doi.org/10.1177/14687941241288178

Bennouna C, Mansourian H, Stark L (2017) Ethical considerations for children's participation in data collection activities during humanitarian emergencies: a Delphi review. Confl Heal 11(1):5. https://doi.org/10.1186/s13031-017-0108-y

Bentele U, Herzog M (2023) Conflict-sensitive research: bringing research ethics into practice. Dev Pract 33(5):656–662. https://doi.org/10.1080/09614524.2023.2213860

Bruno W, Haar RJ (2020) A systematic literature review of the ethics of conducting research in the humanitarian setting. Confl Heal 14(1):27. https://doi.org/10.1186/s13031-020-00282-0
Burke R, Baker S, Molla T et al (2024) How do higher degree research students and supervisors navigate ethics-in-practice for educational research in sensitive or 'fragile'contexts? Br Educ Res J 50(2):837–854. https://doi.org/10.1002/berj.3945
Campbell R, Goodman-Williams R, Javorka M (2019) A trauma-informed approach to sexual violence research ethics and open science. J Interpers Violence 34(23–24):4765–4793. https://doi.org/10.1177/0886260519871530
Chatfield K, Singh M (2025) Research ethics and integrity during pandemics: not unique, but vastly magnified challenges. In: Chatfield K, Singh M (eds) Research ethics and integrity during pandemics: developing the PREPARED Code. SpringerBriefs in Research and Innovation Governance. Springer, Cham, pp 1–7. https://doi.org/10.1007/978-3-031-91324-2
Chiumento A, Khan MN, Rahman A, Frith L (2016) Ethical challenges to research in post-conflict settings. Dev World Bioeth 16:15–28. https://doi.org/10.1111/dewb.12076
Chiumento A, Rahman A, Frith L et al (2017) Ethical standards for mental health and psychosocial support research in emergencies: review of literature and current debates. Glob Health 13(8). https://doi.org/10.1186/s12992-017-0231-y
Chiumento A, Machin L, Rahman A, Frith L (2018) Online interviewing with interpreters in humanitarian contexts. International Journal of Qualitative Studies on Health and Well-Being 13(1). https://doi.org/10.1080/17482631.2018.1444887
Colquhoun HL, Levac D, O'Brien KK et al (2014) Scoping reviews: time for clarity in definition, methods, and reporting. J Clin Epidemiol 67(12):1291–1294. https://doi.org/10.1016/j.jclinepi.2014.03.013
Contractor Q (2016) Fieldwork and social science research ethics. Indian J Med Ethics 1(5):22–23. https://doi.org/10.20529/IJME.2008.008
Council for International Organizations of Medical Sciences (2016) Research in disasters and disease outbreaks. In: International ethical guidelines for health-related research involving humans: Prepared by the Council for International Organizations of Medical Sciences (CIOMS) in collaboration with the World Health Organization (WHO). https://www.ncbi.nlm.nih.gov/books/NBK614401/. Accessed 10 Jan 2026
Cremin H, Aryoubi H, Hajir B et al (2021) Post-abyssal ethics in education research in settings of conflict and crisis: stories from the field. Br Educ Res J 47(4):1102–1119. https://doi.org/10.1002/berj.3712
Davidson N, Hammarberg K, Fisher J (2024) Ethical considerations in research with people from refugee and asylum seeker backgrounds: a systematic review of national and international ethics guidelines. J Bioethical Inq 21(2):261–284. https://doi.org/10.1007/s11673-023-10297-w
Ebubedike M, Akanji T, Kunock AI, Fox A (2023) Ethics for educational research in regions of protracted armed conflict and crisis: a participatory community project in the Lake Chad region. Community Dev J 58(1):102–120. https://doi.org/10.1093/cdj/bsac040
Eriksen C (2017) Research ethics, trauma and self-care: reflections on disaster geographies. Aust Geogr 48(2):273–278. https://doi.org/10.1080/00049182.2016.1230001
Fox A, Baker S, Charitonos K et al (2020) Ethics-in-practice in fragile contexts: research in education for displaced persons, refugees and asylum seekers. Br Educ Res J 46(4):829–847. https://doi.org/10.1002/berj.3618
Franc JM, O'Mathúna D (2025) Navigating research ethics for prehospital and disaster medicine. Prehosp Disaster Med 40(3):125–128. https://doi.org/10.1017/S1049023X25101234
Habib RR (2019) Ethical, methodological, and contextual challenges in research in conflict settings: the case of Syrian refugee children in Lebanon. Confl Heal 13:29. https://doi.org/10.1186/s13031-019-0215-z
Helbardt S, Hellmann-Rajanayagam D, Korff R (2010) War's dark glamour: ethics of research in war and conflict zones. Camb Rev Int Aff 23(2):349–369. https://doi.org/10.1080/09557571003752688

Hunt M, Tansey CM, Anderson J at al (2016) The challenge of timely, responsive and rigorous ethics review of disaster research: views of research ethics committee members. PLoS One 11(6):e0157142 doi:https://doi.org/10.1371/journal.pone.0157142
Hussein G, Elmusharaf K (2019) Mention of ethical review and informed consent in the reports of research undertaken during the armed conflict in Darfur (2004–2012): a systematic review. BMC Med Ethics 20:40. https://doi.org/10.1186/s12910-019-0377-7
Idris I (2019) Doing research in fragile contexts: Literature review. In: GSDRC Research Report no. 1444. GSDRC, University of Birmingham, Birmingham, UK. https://gsdrc.org/wp-content/uploads/2019/04/Doing-Research-in-Fragile-Contexts-Idris-2019.pdf. Accessed 5 Feb 2026
Kohrt BA, Mistry AS, Anand N et al (2019) Health research in humanitarian crises: an urgent global imperative. BMJ Glob Health 4(6):e001870. https://doi.org/10.1136/bmjgh-2019-001870
Kostovicova D, Knott E (2022) Harm, change and unpredictability: the ethics of interviews in conflict research. Qual Res 22(1):56–73. https://doi.org/10.1177/1468794120975657
Krause J (2021) The ethics of ethnographic methods in conflict zones. J Peace Res 58(3):329–341. https://doi.org/10.1177/0022343320971021
Kurihara C, Crawley FP, Baroutsou V et al (2022) The continuation of clinical trials in times of war: a need to develop ethics and situationally adaptive clinical research guidelines. Front Med 9:966220. https://doi.org/10.3389/fmed.2022.966220
Leaning J (2001) Ethics of research in refugee populations. Lancet 357(9266):1432–1433
Levac D, Colquhoun H, O'Brien KK (2010) Scoping studies: advancing the methodology. Implement Sci 5(1):69. https://doi.org/10.1186/1748-5908-5-69
Mezinska S, Kakuk P, Mijaljica G et al (2016) Research in disaster settings: a systematic qualitative review of ethical guidelines. BMC Med Ethics 17:62. https://doi.org/10.1186/s12910-016-0148-7
Munn Z, Peters MDJ, Stern C et al (2018) Systematic review or scoping review? Guidance for authors when choosing between a systematic or scoping review approach. BMC Med Res Methodol 18(1):143. https://doi.org/10.1186/s12874-018-0611-x
National Research Council (2002) Research ethics in complex humanitarian emergencies: summary of a workshop. National Academies Press (US), Washington DC. https://doi.org/10.17226/10481
Ng SH, Kaur S, Cheah PY et al (2024) Migration health ethics in Southeast Asia: a scoping review. Wellcome Open Res 4(8):391. https://doi.org/10.12688/wellcomeopenres.19572.3
O'Mathúna DP (2010) Conducting research in the aftermath of disasters: ethical considerations. J Evid Based Med 3:65–75. https://doi.org/10.1111/j.1756-5391.2010.01076.x
OECD (2022) States of fragility 2022. OECD Publishing, Paris. https://doi.org/10.1787/c7fedf5e-en Accessed 10 Jan 2026
O'Mathúna D (2015) Research ethics in the context of humanitarian emergencies. J Evid Based Med 8:31–35. https://doi.org/10.1111/jebm.12136
O'Mathúna D, Siriwardhana C (2017) Research ethics and evidence for humanitarian health. Lancet 18;390(10109):2228–2229. https://doi.org/10.1016/s0140-6736(17)31276-x
Perakslis ED (2018) Using digital health to enable ethical health research in conflict and other humanitarian settings. Confl Heal 12(1):23. https://doi.org/10.1186/s13031-018-0163-z
Peters MDJ, Godfrey C, P MI et al (2020) Scoping reviews. In: Aromataris E, Lockwood C, Porritt K et al (eds) JBI Manual for Evidence Synthesis. https://doi.org/10.46658/JBIMES-24-09
Pincock K, Bakunzi W (2021) Power, participation, and 'peer researchers': addressing gaps in refugee research ethics guidance. J Refug Stud 34(2):2333–2348. https://doi.org/10.1093/jrs/feaa060
Poole DN, Raymond NA, Berens J et al (2021) A combination sampling approach for epidemiologic research in humanitarian settings: a case analysis of a study of depressive disorder prevalence among refugees in Greece. BMC Public Health 21:272. https://doi.org/10.1186/s12889-021-10342-6
PREPARED (2025a) The PREPARED code—a global code of conduct for research during pandemics. https://prepared-project.eu/prepared-code/. Accessed 10 Oct 2025

PREPARED (2025b) The TRUST Code Supplement—a global code of conduct for research in fragile settings. https://www.globalcodeofconduct.org/trust-code-supplement/. Accessed 11 Dec 2025

Refsdal NO (2014) Experiences from coordinating research after the 2011 terrorist attacks in Norway. Eur J Psychotraumatol 5(1). https://doi.org/10.3402/ejpt.v5.23215

Schiltz J, Büscher K (2018) Brokering research with war-affected people: the tense relationship between opportunities and ethics. Ethnography 19(1):124–146. https://doi.org/10.1177/1466138116664542

Schopper D, Upshur R, Matthys F et al (2009) Research ethics review in humanitarian contexts: the experience of the independent ethics review board of Médecins Sans Frontières. PLoS Med 6(7):e1000115. https://doi.org/10.1371/journal.pmed.1000115

Seagle EE, Dam AJ, Shah PP et al (2020) Research ethics and refugee health: a review of reported considerations and applications in published refugee health literature, 2015–2018. Confl Heal 14(1):39. https://doi.org/10.1186/s13031-020-00283-z

Sen K, Hussain H, Faisal WA (2016) Ethics in times of conflict: some reflections on Syria, in the backdrop of Iraq: BMJ. Glob Health 1:e000149. https://gh.bmj.com/content/1/3/e000149

Sombié I, Aidam J, Montorzi G (2017) Evaluation of regional project to strengthen national health research systems in four countries in West Africa: lessons learned. Health Res Policy Syst 15(Suppl 1):46. https://doi.org/10.1186/s12961-017-0214-8

Stratton SJ (2012) Ethics in EMS and disaster research. Prehosp Disaster Med 27(06):495. https://doi.org/10.1017/S1049023X12001495

Tansey CM, Anderson J, Boulanger RF et al (2017) Familiar ethical issues amplified: how members of research ethics committees describe ethical distinctions between disaster and non-disaster research. BMC Med Ethics 18(1):44. https://doi.org/10.1186/s12910-017-0203-z

Thomas SN, Weber S, Bradbury-Jones C (2022) Using participatory and creative methods to research gender-based violence in the Global South and with Indigenous communities: findings from a scoping review. Trauma Violence Abuse 23(2):342–355. https://doi.org/10.1177/1524838020925775

Thomson B, Mehta S, Robinson C (2024) Scoping review and thematic analysis of informed consent in humanitarian emergencies. BMC Med Ethics 25:135. https://doi.org/10.1186/s12910-024-01125-w

Tiruneh G, Yilma M, Wakuma B et al (2024) Compliance with research ethics in epidemiological studies targeted to conflict-affected areas in Western Ethiopia: validity of informed consent (VIC) by information comprehension and voluntariness (ICV). BMC Med Ethics 25(1):9. https://doi.org/10.1186/s12910-024-01003-5

Van der Loos KI, Longstaff H, Virani A, Illes J (2014) Consent in escrow. J Law Biosci 2(1):69–78. https://doi.org/10.1093/jlb/lsu031

Vervliet M, Rousseau C, Broekaert E, Derluyn I (2015) Multilayered ethics in research involving unaccompanied refugee minors. J Refug Stud 28(4):468–485. https://doi.org/10.1093/jrs/feu039

Vogel B (2025) Mind the ethics gap: embedding research ethics into student fieldtrips to conflict and development settings. Politics 45(4):655–664. https://doi.org/10.1177/02633957251344550

Vogel C, Musamba J (2025) Towards a politics of collaborative worldmaking: ethics, epistemologies and mutual positionalities in conflict research. Ethnography 26(3):603–623. https://doi.org/10.1177/14661381221090895

WHO (2020) Guidance for research ethics committees for rapid review of research during public health emergencies. World Health. Organization. https://www.who.int/publications/i/item/9789240006218. Accessed 7 Feb 2026

World Conferences on Research Integrity (2010) Singapore statement on research integrity. https://www.wcrif.org/guidance/singapore-statement. Accessed 7 Feb 2026

Xi Y, Chen R, Gillespie AL et al (2019) Mental health workers perceptions of disaster response in China. BMC Public Health 19(1):11. https://doi.org/10.1186/s12889-018-6313-9

Chapter 4
Research Ethics and Integrity Challenges in Fragile Settings: A Scoping Review–Informed Evidence Mapping of German Sources

Abstract Fragile settings are characterised by the presence of significant risks, combined with an insufficient capacity to manage or mitigate those risks. Researchers in these settings can encounter a wide range of research ethics and integrity challenges. This chapter presents a scoping review–informed evidence mapping of German-language literature that identifies the specific ethical challenges associated with conducting research in fragile settings. The review identified 16 major challenges: special burdens research participants face before the research; health, wellbeing and safety concerns for research participants; negative impacts arising from disingenuous or patronising research; the potential for direct harm caused by researchers; unrealistic expectations of research participants; retraumatisation risks for research participants; the protection-inclusion dilemma; the humanitarian misconception; payment-related challenges; misuse of research data and findings; health, wellbeing and safety concerns for researchers and support teams; ethics governance challenges; time-related pressures; conflicts of interest; challenges for the informed consent process; and issues related to data reliability.

Keywords Global research ethics · Fragile settings · Humanitarian misconception · Patronising research

4.1 Introduction

The OECD (2022) defines fragile settings as environments where significant risks intersect with an insufficient coping capacity to manage or mitigate those risks. This situation can lead to "negative outcomes, including violence, poverty, inequality, displacement, and environmental and political degradation" (ibid).

World Vision, a global relief, development and advocacy organisation, defines fragile settings as: "The world's most dangerous places … fraught with chronic instability, conflict and violence, trapping large numbers of people in a cycle of desperation and poverty" (Ramos 2025).

D. Schroeder et al., *Research Ethics and Integrity in Fragile Settings*, SpringerBriefs in Research and Innovation Governance,
https://doi.org/10.1007/978-3-032-36141-7_4

To ensure that important research can be undertaken ethically, a team was funded by the European Commission (PREPARED n.d.) to develop ethics codes for global crises. The first code that was developed focused on pandemics (Chatfield and Singh 2025). The second code focuses on research in fragile settings.

To ensure that the fragile settings code was firmly grounded in practical realities, it was important to capture as many real-world challenges across different types of research and contextual environments as possible.

To this end, the team undertook three studies:

1. a focused scoping review of relevant academic literature published in English (see Chap. 3)
2. a scoping review–informed evidence mapping of relevant academic literature published in German
3. a survey to investigate the experiences and views of people who have conducted research in fragile settings—an anonymous electronic survey, gathering data from 37 researchers to reveal the risks and challenges to ethics and integrity that they had personally experienced or observed in the field (see Chap. 5).

This chapter is about the scoping review–informed evidence mapping of relevant academic literature published in German.

4.2 Methods

4.2.1 Method Selection

While science is international, "scientific publishing is dominated by English-language publications. This disproportionately benefits native or fluent English speakers" (Nature 2023). We chose German as a second language for evidence mapping for three reasons. First, and pragmatically, the necessary research skills combined with a native command of German were available within the existing team. Second, German is historically a highly prominent language in scientific discourse (Porzucki 2014). Third, German will "continue to be of great importance in the humanities and social sciences in the future" (Pfänder 2021).

Scoping reviews are often selected over other types of reviews (like systematic reviews, narrative reviews or meta-analyses) when the primary purpose of the study is to explore the breadth of existing evidence and map key concepts rather than address a narrowly defined question about effectiveness or causality (Peters et al. 2020, 2021). The scoping review approach is considered particularly appropriate where comprehensive but flexible mapping is needed rather than detailed critical appraisal, or when a topic is emerging. (Arksey and O'Malley 2005; Munn et al. 2018).

A scoping review is typically completed in 9–12 months, but can take more than a year. As we only had 6 months, we followed the five-stage framework of Arksey and

O'Malley (2005) for scoping reviews, with adaptions based on the interdisciplinary and cross-language nature of the topic. Searches were conducted in a single primary database (Google Scholar). While our search strategy was not multidatabase, the overall methodological structure—including a broad research question, predefined search terms, explicit inclusion/exclusion criteria, dual-reviewer screening and structured data charting—aligns with a scoping review methodology and is consistent with established evidence-mapping approaches.

In addition, there are scholars who argue that one-database searches can be sufficient for major evidence synthesis (van Enst et al. 2014; Rice et al. 2016), and some one-database scoping reviews have been published (Hearn et al. 2019; Hopia and Heikkilä 2019; Billones et al. 2021).

The five steps of a scoping review proposed by Arksey and O'Malley (2005) were adopted for this review, augmented by a sixth step as suggested by Mak and Thomas (2022).

- Step 1. Identifying the research question
- Step 2. Identifying relevant studies
- Step 3. Study selection
- Step 4. Charting the data
- Step 5. Collating, summarising and reporting the results
- Step 6. Consulting stakeholders (optional)

4.2.2 Step 1: Identifying the Research Question

The German review adopted the same broad research question as the English review, namely: *What are the challenges for research ethics and research integrity when undertaking research in fragile settings?*

Which translated as:

> Was sind die forschungsethischen und integritätsbezogenen Herausforderungen für die Forschung in fragilen Gebieten?

Also, like the English review, the objectives of the German scoping review were:

1. Identify specific examples of research ethics issues that arise and/or are exacerbated when research is conducted in fragile settings.
2. Identify specific examples of research integrity issues that arise and/or are exacerbated when research is conducted in fragile settings.
3. Theme the identified examples according to the nature and type of research ethics or research integrity challenge that they represent.

4.2.3 Step 2: Identifying Relevant Studies

The search strategy was adapted to the context of German academic publishing, where Google Scholar is the academic search engine of choice.

In alignment with the English language review (see Chap. 3), the search was conducted with the terms "Forschungsethik" (research ethics) OR "Forschungsintegrität" (research integrity) AND "type of fragile setting".

As "fragile setting" could include a very broad range of settings, to ensure fit with the study objectives we specified the following types of fragile settings:

1. conflict or post-conflict settings
2. settings that had recently experienced natural disasters such as earthquakes, tsunamis, hurricanes or floods
3. settings that were or had been devastated by climate change
4. refugee camps in lower-income regions
5. settings characterised by high violence and crime.

All five settings create significant risks for those who live and work there, making them highly dangerous places, including for conducting research (Ramos 2025).

Two of the terms used for the review do not relate directly to the identified settings but are supplementary: first, the term "humanitär", possibly leading to results on humanitarian interventions in settings that we had not considered, and second, the term "fragil", possibly leading to results on fragile settings that we also had not considered. Thus the search included nine separate search terms to capture the breadth of types of fragile settings as shown in Table 4.1.

Rather than one search combining all types of fragile settings, nine separate searches were conducted. The searches were performed without date restrictions.

Table 4.1 German search terms

Term "Forschungsethik" OR "Forschungsintegrität" AND	
German search term	Type of setting
1. "Konflikt"	Conflict or post-conflict
2. "Disaster" OR "Katastrophe"	Experienced natural disasters
3. "Erdbeben" OR "Tsunami" OR "Hurrikan" OR "Überschwemmung"	Experienced earthquakes, tsunamis, hurricanes or floods
4. "Krieg" OR "bewaffnet" OR "militä*" OR "Aufstand"	Experienced war, military interventions, revolts or riots
5. "Flüchtling" OR "Flucht"	Refugees are present
6. "hohe Kriminalität" OR "Gewalt"	Characterised by high violence and crime
7. "Klimawandel"	Devastated by climate change
8. "fragil"	Characterised as "fragile"
9. "Humanitär"	Seeing humanitarian intervention

Table 4.2 Initial returns for the nine searches in Google Scholar

Term "Forschungsethik" OR "Forschungsintegrität" AND	Results
1. "Konflikt"	1330
2. "Humanitär"	25
3. "Disaster" OR "Katastrophe"	289
4. "Erdbeben" OR "Tsunami" OR "Hurrikan" OR "Überschwemmung"	88
5. "Krieg" OR "bewaffnet" OR "militä*" OR "Aufstand"	779
6. "Flüchtling" OR "Flucht"	683
7. "hohe Kriminalität" OR "Gewalt"	1520
8. "fragil"	82
9. "Klimawandel"	257
Total results	5053

As is common with searches in Google Scholar, some of the searches returned a high number of hits, as indicated in Table 4.2.

4.2.4 Step 3: Selecting Studies to Be Included in the Review

For each of the nine searches, the results were sorted in Google Scholar by relevance and the titles and abstracts were screened by researcher DS, reading abstracts where available or selections from the full text. Each result was assessed for inclusion against the inclusion and exclusion criteria (Table 4.3).

The process of assessing relevance was continued until the point of saturation, when no more relevant articles appeared. From that point, a further eight to ten entries were examined to be sure that the "not relevant" point had been reliably identified, bringing the overall figure of publications assessed to 222. Researcher LT then checked the "not useful" assignation by DS, rather than performing the same task. Discrepancies were discussed and resolved in a meeting.

Out of a possible 5053 publications, 139 were identified as having content that was in some way relevant to research ethics and integrity in fragile settings. The full citations and abstracts (where available) of the 139 publications were extracted to an Excel spreadsheet.

While the initial screening allowed content related to research ethics and integrity to be identified, it could not go as far as determining the relevance of this material to the study objectives. For instance, some articles addressed only matters related to research ethics approval, and not ethics issues involving the conduct of research in a fragile setting. To identify materials that were relevant to the review objectives, further screening was required.

Hence, DS read all 139 publications to assess their relevance to the review objectives. A new column was added to the Excel file, noting reasons for including or

Table 4.3 Inclusion and exclusion criteria

Criteria	Inclusion	Exclusion
Language	German	Non-German
Type of publication	All published literature types	
Nature of content	Based upon experiences of research in fragile settings, either first or second hand	Purely theoretical, not based on real-world experiences
Research ethics and integrity content	Some focus on particular research ethics and/or research integrity issues	Brief mention of research ethics and/or research integrity (e.g., in references)
Conflict-related materials*	Research conducted in conflict or post-conflict regions	Conflicts of interest not related to fragile settings
Refugee research setting	Low-income settings	High-income settings**

* The Google Scholar search related to "Konflikt" returned many publications that described conflicts of interest rather than conflict within the research setting. Some conflicts of interest arise or are exacerbated in fragile settings and examples of these were included. Where conflicts of interest were not related to a fragile setting context, these examples were excluded

* Research involving refugees who were residing in a high-income country when the research was conducted was excluded from this study. While research involving refugees in high-income countries can be fraught with ethical difficulties (Kabranian-Melkonian 2015), these are generally of a different nature to those encountered in fragile settings, as the refugees are—technically, at least—in a safe place,[1] not a place that is characterised by, for instance, armed conflict

[1] We understand that settings that are safe generally can be unsafe for refugees specifically (see Pinillos-Franco and Kawachi 2022). However, the challenges this presents are different from those in settings where everybody is unsafe

excluding each of the 139 works from the review. Publications were colour-coded according to their assessment as "relevant", "irrelevant" or "unclear".

LT then undertook a check by examining 100% of the "unclear" publications and 40% of those coded "relevant" or "irrelevant". Agreement was reached through discussion on the "unclear" publications. In three cases, a "relevant" was changed to "irrelevant" after discussion, leading to the results summarised in Table 4.4.

Thus the search produced 12 articles or books that met the inclusion criteria and were relevant to the review objectives. These publications are listed in Table 4.5 and Fig. 4.1 provides a selection flow chart.

4.2.5 *Step 4: Charting the Data*

Given the heterogeneity of the publications identified, neither the extraction of methods nor an assessment of scientific quality would be meaningful in this review. Instead, the scoping review approach described by Arksey and O'Malley (2005) and by Mak and Thomas (2022) was modified for the very specific purpose of informing

Table 4.4 Relevant articles or books per search term

Term "Forschungsethik" OR "Forschungsintegrität" AND	Results
1. "Konflikt"	2/10
2. "Humanitär"	2/13
3. "Disaster" OR "Katastrophe"	0/14
4. "Erdbeben" OR "Tsunami" OR "Hurrikan" OR "Überschwemmung"	4/35
5. "Krieg" OR "bewaffnet" OR "militä*" OR "Aufstand"	1/15
6. "Flüchtling" OR "Flucht"	2/19
7. "hohe Kriminalität" OR "Gewalt"	0/15
8. "fragil"	1/9
9. "Klimawandel"	0/9
Total results	12/139

the development of a research ethics and integrity code. This required the identification and extraction of examples of research ethics and integrity issues that arise or are exacerbated when research is conducted in fragile settings (objectives 1 and 2).

For this task, DS extracted relevant quotations from the 12 publications to an Excel file. Where the authors of a work repeated their discussion of particular research ethics and integrity issues, only one quotation was extracted. Each quotation was then summarised to identify a specific issue.

LT quality-checked the screening and coding decisions by reviewing 10% of the quotations marked as relevant and 10% of those marked as irrelevant, recording whether she agreed or disagreed and adding her reasons. She also assessed 100% of the entries marked as unclear, providing her own judgement in each case. This ensured consistency in the inclusion of quotations.

In the second step, LT read the two most content-rich publications, Krause and Williams (2021) and Dittmer and Lorenz (2018), reviewed all related quotations, and confirmed that no additions or deletions were needed, as any potential additions would duplicate content already extracted.

Following collaborative work and discussions, DS and LT identified 94 research ethics or research integrity issues.

4.2.6 Step 5: Collating, Summarising and Reporting the Results

The next step in a scoping review normally involves thematic analysis to collate, summarise and report the results (Mak and Thomas 2022). There are many different approaches that can be taken to the theming stage, but for this goal-oriented review, theming was aligned with objective 3:

> Theme the identified examples according to the nature and type of research ethics or research integrity challenge that they represent.

Table 4.5 Relevant articles or books

Relevant articles or books	Type
Abdelkader S, Bach M, Narawitz L et al. (2021) Forschungsethik als politische Verantwortung: Reflexionen zum Veröffentlichen in der Flucht*Migrationsforschung. In: Bach M, Narawitz L, Schroeder J et al. (eds) FluchtMigrationsForschung im Widerstreit. Waxmann, Münster, pp. 211–225	Academic book chapter
Buckley-Zistel S (2012) "Ich bin dann mal weg." Feldforschung im Studium der Friedens- und Konfliktforschung. Zeitschrift für Friedens- und Konfliktforschung 1(2):315–323	Peer-reviewed journal article
Dittmer C, Lorenz DF (2018) Forschen im Kontext von Vulnerabilität und extremem Leid: ethische Fragen der sozialwissenschaftlichen Katastrophenforschung. Forum Qualitative Sozialforschung / Forum: Qualitative Social Research 19(3)	Peer-reviewed journal article
Erdmann N, Bartmann S, Haefker M et al. (2022) Epilog: Erkenntnisse und Perspektiven für gesellschaftlich wahrgenommene Ausnahmesituationen. In: Bartmann S, Erdmann N, Haefker M et al. (eds) Verstehendes Forschen in der Pandemie und anderen Ausnahmesituationen. Verlag Barbara Budrich, pp. 159–168	Academic book chapter
Exo M (2017) das übergangene Wissen: Eine dekoloniale Kritik des liberalen peacebuilding durch basispolitische Organisationen in Afghanistan. Transcript Verlag, Bielefeld	Academic book
Krause U (2016) Ethische Überlegungen zur Feldforschung: Impulse für die Untersuchung konfliktbedingter Flucht. CSS working paper no 20. Zentrum für Konfliktforschung der Philipps-Universität Marburg	Academic working paper
Krause U, Williams T (2021). Flexible Ethikgremien: Impulse zur Institutionalisierung ethisch verantwortlicher Feldforschung in der Konflikt- und Fluchtforschung. Soziale Probleme 32:97–113	Peer-reviewed journal article
Niemann H, Schröder U (2020) Ein Feld in Bewegung: Friedens- und Konfliktforschung in herausfordernden Zeiten. Zeitschrift für Friedens- und Konfliktforschung 9:131–142	Peer-reviewed journal article
Schrader H (2024) Ostwärts! Lehrforschungen der Jahre 1998–2023 in fremdkulturellen Kontexten. DSS Working Paper Nr. 83. Institut für Gesellschaftswissenschaften, Otto-von-Guericke-Universität Magdeburg	Academic working paper
Schwenken H (2015) Rechtlos, aber nicht ohne Stimme: Politische Mobilisierungen um irreguläre migration in die Europäische union. Transcript Verlag, Bielefeld	Academic book
Sieber Egger A (2011) Krieg im Frieden: Frauen in Bosnien-Herzegowina und ihr Umgang mit der Vergangenheit. Transcript Verlag, Bielefeld	Academic book
Włostowski Ł (2024) camp als commons? Ein selbstverwaltetes Projekt von Geflüchteten in Lavrio (Griechenland). Alice Salomon Hochschule Berlin	MA disser-tation

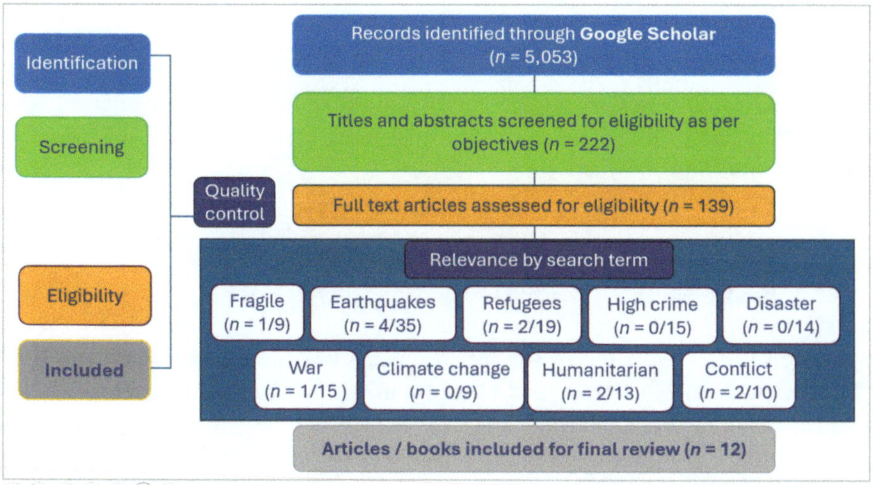

Fig. 4.1 Selection flow chart

Thus the 94 research ethics and integrity issues were themed according to the nature and type of research ethics or integrity challenge that they represented. This was undertaken by DS first and then checked by LT. The theming led to the identification of 16 challenges for research ethics and/or research integrity. These challenges, listed below, are grouped below according to who is mostly affected, even though challenges can be applicable to more than one group.

Challenges for Research Participants

1. special burdens research participants face before the research
2. health, wellbeing and safety concerns for research participants
3. negative impacts arising from disingenuous or patronising research
4. the potential for direct harm caused by researchers
5. unrealistic expectations of research participants
6. retraumatisation risks for research participants
7. the protection-inclusion dilemma
8. the humanitarian misconception
9. payment-related challenges
10. misuse of research data and findings.

Challenges for Researchers

1. health, wellbeing and safety concerns for researchers and support teams
2. ethics governance challenges
3. time-related pressures
4. conflicts of interest

5. challenges for the informed consent process
6. issues related to data reliability.

Each of these challenges is described briefly below and illustrated by selected ethics and integrity issues identified in the German literature. (All quotations have been translated into English by DS.)

4.2.6.1 Special Burdens Research Participants Face Before the Research

In fragile settings, potential research participants will have experienced and might still be experiencing violence and other traumatic burdens (Krause and Williams 2021). Research participants who are refugees may also carry the burden of knowing that they might be imprisoned or even tortured or sentenced to death in their country of origin (Włostowski 2024).

And the challenges go beyond the immediate burdens, such as war or natural disasters. Half of those who experience rape, displacement, war or torture are left traumatised when the conflict situation is over (Krause 2016). Sieber Egger explains that even years after the war in Bosnia ended, research participants in her study still suffered from anxiety, restlessness, massive sleep disorders and recurrent nightmares, lacking a safe and positive perspective on the future (Sieber Egger 2011: 74).

These pre-existing, special burdens pose a profound ethical challenge for research in fragile settings because participation in research may risk adding to or exacerbating them. As Buckley-Zistel (2012) notes, such contexts compel researchers to reflect critically on whether, when, and how research can be conducted without placing additional burdens on those who have already suffered extensive loss.

4.2.6.2 Health, Wellbeing and Safety Concerns for Research Participants

In fragile settings, potential research participants are highly likely to face serious health, wellbeing and safety threats, even without the presence of researchers. Yet research activities can aggravate these threats. Types of harms that may be encountered through involvement in research in fragile settings can be diverse, including, for instance, physical attacks (Buckley-Zistel 2012) and harms to societal or political station (ibid).

Researchers can cause or aggravate health and safety concerns if they prioritise the hunt for data over the safety of research participants (Krause and Williams 2021). And an overly strong focus on data acquisition might lead them to ignore the needs of research participants (Krause 2016). For instance, Krause and Williams (2021) refer to incoming researchers demanding, “Take us to the raped women!” in the process of uncovering the extent of violence experienced (see also Sect. 4.2.6.6). Another example of risk-enhancing behaviour on the part of researchers is the inclusion in their outputs of photographs and names that identify research participants (Krause 2016).

In some cases health, wellbeing and safety concerns might apply only to subgroups of potential research participants. For instance, in Uganda, whose Anti-Homosexuality Act was introduced in 2023, LGBTQ+ groups involved in research can be targeted and blamed for violence inflicted upon them (Krause and Williams 2021). Equally, a biased or careless selection of research participants—for instance, from only one ethnic group—can lead to safety concerns for the chosen group (ibid).

Generally, all potential research participants can be at risk when authorities in the location object to the research and contemplate penalising those who take part (Krause and Williams 2021).

4.2.6.3 Negative Impacts Arising from Disingenuous or Patronising Research

Negative impacts on research participants may extend beyond concerns related to health, wellbeing and safety. In particular, disingenuous or patronising conduct by researchers can cause significant harm.

Disingenuous conduct includes making promises that cannot be fulfilled: for example, promising feedback on research findings but failing to provide it, or when they engage briefly with people present in refugee camps and conflict areas without considering the longer-term consequences of their presence (Krause 2016).

Patronising conduct has also been termed "epistemic violence" (Niemann and Schroeder 2020), under which local, experiential and nonacademic expertise is not acknowledged as legitimate or valuable (ibid). Instead, local knowledge systems may be marginalised or downgraded throughout the research process (ibid). In contrast, nonpatronising research would see researchers recognising the limits of their own knowledge, especially in conflict research, and integrate Indigenous and traditional knowledge meaningfully into the entire research process (ibid).

Power imbalances further exacerbate these risks if they restrict opportunities for research participants to recount their full stories. Researchers who adopt patronising attitudes may fail to create the safe space that research participants need to provide their perspectives (Krause and Williams 2021). Such disingenuous and patronising interactions can ultimately generate anxiety, frustration and feelings of disrespect among research participants (ibid).

4.2.6.4 The Potential for Direct Harm Caused by Researchers

In the worst cases, researchers cause harm simply through their presence. For instance, researchers might hamper rescue efforts (Dittmer and Lorenz 2018), take up valuable infrastructure that rescuers could use (ibid), or even engage in war tourism (Buckley-Zistel 2012).

4.2.6.5 Unrealistic Expectations of Research Participants

Unrealistic expectations among research participants are a significant concern for responsible researchers in fragile settings. Research participants may, for example, hope that research outcomes will help them make sense of the terrible events they have endured (Dittmer and Lorenz 2018). While researchers may be motivated by a genuine desire to do good, they are often unable to meet such expectations, particularly when participants' hopes exceed what research can realistically deliver (ibid).

Furthermore, the hopes of research funders and those of research participants might not coincide, reducing the likelihood that the hopes of research participants will be fulfilled (ibid) (see also Sect. 4.2.6.14 below). Without careful and honest expectation management, research may result in disappointment and emotional harm for research participants (ibid) (see also Sect 4.2.6.8).

4.2.6.6 Retraumatisation Risks for Research Participants

There is another negative impact on research participants that is significant enough to warrant separating out from the impacts mentioned above, and that is the possibility of retraumatising them.

Inviting research participants in fragile settings, many of whom have personally experienced violence, to recount their experiences carries a significant risk of retraumatization (Krause and Williams 2021). In particular, biography-based or narrative interview methods pose a potential for retraumatisation (Sieber Egger 2011: 73) as they may require participants to revisit distressing events. Even merely approaching research participants for informed consent in fragile settings, especially disaster settings, can create harm when they do not want to be reminded of the disaster (Dittmer and Lorenz 2018). However, the informed consent process must communicate details about the research topic *before* agreement is obtained from the potential study participant (ibid).

At the same time, researchers must be aware that some potential research participants may have a strong desire to talk about their experiences of suffering (ibid: 74). Accordingly, overly protective assumptions about the risk of retraumatisation can also be inappropriate and may themselves cause harm by silencing participants or denying them agency (Dittmer and Lorenz 2018). This can lead to a protection-inclusion dilemma as described in the next section.

4.2.6.7 The Protection-Inclusion Dilemma

One governance approach to research in fragile settings is to assume that all such research is ethically too sensitive and that potential participants are simply too vulnerable to take part in research. For instance, since the 11 September 2001 attacks in the United States, research ethics committees have regularly deemed disaster victims too vulnerable for involvement in research (Dittmer and Lorenz 2018). However, this approach raises the protection-inclusion dilemma (Schroeder et al. 2024). Do you exclude potential research participants to protect them from research? Or do you include them so as not to appear patronising?

On the one hand, it has been argued that refugees or disaster-affected people deserve special protection due to their marginalised position (ibid). On the other hand, this can be considered a patronising approach that labels those involved as "victims" and limits the agency of people who might want to contribute to research (ibid). As noted in Sect. 4.2.6.6, *not* obtaining the views of disaster survivors can also create harm (Sieber Egger 2011: 73). A research participant may feel impelled to let the world know what happened to them (ibid: 74).

4.2.6.8 The Humanitarian Misconception

In medical research, one speaks of a "therapeutic misconception", when research participants fail to realise that they are unlikely to obtain therapeutic benefits from an experimental study (Lidz et al. 2004).

The equivalent for research in fragile settings is a humanitarian misconception. First, it may be unclear to potential participants whether visitors to the area are researchers, aid workers or both (Dittmer and Lorenz 2018). If they are researchers without any means to provide aid, research participants who are unaware of this distinction might take part under a mistaken assumption. Second, researchers working in refugee camps might be mistaken for personnel conducting asylum-decision interviews (Krause 2016), again potentially gaining cooperation from research participants on a false premise. This is in conflict with the requirement that research must always be undertaken with volunteers.

4.2.6.9 Payment-Related Challenges

Paying people in vulnerable situations for research participation can create safety issues for them, as the financial incentive might persuade them to take part in research against their better judgement (Krause and Williams 2021). This is called "undue inducement" in the research ethics literature (Phillips 2023).

As a result, and to avoid these safety difficulties, some researchers do not provide payment, but only feedback and photographs (Schrader 2024). This may also happen because research funders do not normally provide money to pay research participants (ibid) (Krause 2016).

Additionally, payment can distort research results, because monetary compensation can lead to research participants feeling obliged to offer specific answers (ibid).

4.2.6.10 Misuse of Research Data and Findings

The risk that research data or research results will be misused, to the detriment of research participants and their communities, is high for research in fragile settings. For instance, insights into migration might lead to new, repressive measures against migrants (Schwenken 2015: 20; Abdelkader et al. 2021). Or research results might increase local conflicts (Dittmer and Lorenz 2018).

At the same time, researchers have little or no influence over how their results are used (Abdelkader et al. 2021) in potentially highly politicised contexts. Schwenken (2015: 19) therefore avoids using language in her reports that could increase conflict, such as "illegal immigrants", because this term suggests that the individuals have done something wrong.

4.2.6.11 Health, Wellbeing and Safety Concerns for Researchers and Support Teams

Some risks, in particular concerns for health, wellbeing and safety, are not unique to research participants; they also apply to researchers and support teams. Diseases, unsafe transport and physical attacks are examples of health and safety concerns that researchers can face (Buckley-Zistel 2012; Krause and Williams 2021). The dangers, according to the German literature, are higher for researchers from the LGBTQ+ community than for others (Krause and Williams 2021).

However, what is particularly noteworthy is that researchers themselves can be traumatised, especially if they have not been involved in disaster research before (Dittmer and Lorenz 2018). This applies also to research interpreters, who can suffer harm as they communicate the accounts of those who suffer (ibid).

Feelings of helplessness are also features of the situation for researchers in fragile settings, both when they want to help but cannot or should not (e.g., because they are required to stay neutral) (ibid) and when they come face to face with traumatised people (Buckley-Zistel 2012). In either case, they might feel uncertain about how to engage, or even powerless to do so at all.

4.2.6.12 Ethics Governance Challenges

Research conducted in fragile settings is inherently sensitive, yet it often receives less institutional and governance support than many other undertakings. For instance, there is little discussion in peace and conflict research about research ethics (Krause 2016). In Germany, there are no research ethics guidelines that German researchers working in conflict research can use (Krause and Williams 2021). Some social scientists working in fragile settings do not even have access to a research ethics committee (REC) (ibid), which also makes publication in English language journals difficult, as those publications normally require proof of ethics approval (ibid).

At the same time, those who have the support of a REC might face other problems, as described in Chap. 2. According to the literature, RECs in German-speaking countries (in most cases far away from the research location) might not be best placed to decide, for instance, whether research will hamper rescue efforts or what the potential harm for research participants could be (Dittmer and Lorenz 2018). In addition, RECs might give the avoidance of harm to the institution priority over the avoidance of harm to the research participants (ibid). Some researchers, it is argued, might even manipulate a study protocol to avoid rejection by their REC (Krause and Williams 2021).

One example of standard REC practice potentially being in conflict with ethical conduct in the field is that of interviews with traumatised people. Standard protocol might require the interview questions, at least in outline, to be submitted to the REC during the ethics approval process, but this could be unethical if it forces participants who have experienced trauma into specific directions (Sieben Egger 2014: 75).

4.2.6.13 Time-Related Pressures

Even where governance mechanisms exist, pressures of time present a challenge that is magnified for research in fragile settings, for two main reasons. First, researchers may feel the need to arrive in the field with the minimum of delay, especially for disaster research. This means there may be less time than usual to prepare a protocol for a REC (Dittmer and Lorenz 2018). Second (as illustrated in Chap. 2), RECs may be especially cautious when assessing research that is to be conducted in a fragile setting, and this caution may lead to delays in approvals (ibid). As disasters cannot be predicted, it is not possible to prepare research ethics protocols in advance, and generic study protocols have not proven useful in disaster research (ibid).

4.2.6.14 Conflicts of Interest

Conflicts of interest can arise, especially in peace and conflict research, because such studies are almost always close to politics (Niemann and Schröder 2020). When a dependence on project funding from external sources is added to the mix, the potential for conflicts of interest is even higher (ibid). The proximity to politics and dependence upon external funding can also create conflicts of interest between the expectations of funders and those of research participants (Dittmer and Lorenz 2018), as noted in Sect. 4.2.6.5 in relation to unrealistic expectations. For instance, research participants might expect research results to improve the situation in their refugee camp, while a funder might have a different agenda (ibid), such as reducing numbers in the camp.

In the worst cases, researchers who want to gain timely access to disaster or conflict areas can only do so by being, in effect, complicit with authoritarian authorities to a degree that borders on corruption (Dittmer and Lorenz 2018).

4.2.6.15 Challenges for the Informed Consent Process

Once researchers are in the field, their challenges continue with difficulties in applying standard procedures for informed consent. As Dittmer and Lorenz (2018) note, linguistic differences in cross-cultural research in fragile settings can be difficult to navigate. For instance, one project in the Himalayas required interpreters in four languages, and another in the same area involved pilgrims to whom one could not talk about nonspiritual matters (ibid). Hence, the mere process of informed consent was blocked by this cultural impediment, even though the pilgrims might have liked to contribute to research that would improve the safety of future pilgrims.

Another challenge noted in the German literature was illiteracy among research participants, which was thought to occur more frequently in fragile than in other settings (Krause 2016).

4.2.6.16 Issues Related to Data Reliability

While the German literature only mentioned issues related to data reliability for research in fragile settings once (Buckley-Zistel 2012), it is nevertheless an important concern.

4.2.7 Step 6: Consulting Stakeholders (Optional)

While Arksey and O'Malley (2005) do not anticipate a consultation step, Mak and Thomas (2022) do. They suggest gaining additional insights and shedding light on the review's findings as reasons for stakeholder consultation. Consulting stakeholders might, in particular, identify gaps in the review.

> While a stakeholder consultation has been named as the final step of a review, it can be incorporated throughout the review stages and can occur through focus groups, individual interviews, or surveys. (ibid)

Two types of stakeholder consultations were undertaken to refine the review results. First, early findings from the English and the German reviews were validated through a survey (see Chap. 5). Second, once the findings of all three processes (two reviews, one survey) were condensed into an ethics code—namely, the TRUST Code Supplement—the code was opened to broad consultation. For instance, 40 experts on research ethics and integrity, as well as representatives from marginalised populations, provided consultative input, which led to refinements of the code (see Fig. 4.2). After these refinements, a public, web-based consultation was opened for 2 months, which also led to further refinements.

Fig. 4.2 Consultations on the TRUST Code Supplement at the UNESCO conference in June 2025

4.3 Discussion

The main purpose of the scoping review–informed evidence mapping of German sources on research ethics and integrity challenges in fragile settings was to inform a research ethics code. To ensure that the code was grounded in real-world challenges and not in the drafters' views about what those challenges *might* be, two reviews and one survey were undertaken. After those three steps, an ethics code was drafted, and stakeholders were invited to help refine the code.

Three main research questions were answered with this review. We found specific examples of *research ethics issues* that arise and/or are exacerbated when research is conducted in fragile settings, and specific examples of *research integrity issues* that arise and/or are exacerbated when research is conducted in fragile settings, and we themed the identified examples.

Sixteen groups of issues or challenges were identified, with the overwhelming majority falling into the area of research ethics. Only one group of issues, namely unreliable data, fell into the area of research integrity.[1]

The challenges for research participants were more numerous than those for researchers (10 out of 16). This is unsurprising, as researchers, especially those who come from outside the fragile setting, can choose to abandon the research and return to safety.

Research can be an additional burden that impacts negatively on the health, well-being and safety of research participants. Researchers can create direct harm by impeding rescue operations or acting disingenuously or patronisingly. They can

[1] Patronising conduct by researchers is mainly grouped with harm to research participants, even though it is also unlikely to produce excellent research.

retraumatise research participants or be received like aid workers, thereby potentially enrolling participants (deliberately or not) under false pretences. Most importantly, their results can be misused to the detriment of research participants and their communities, for instance to justify more repressive migration rules.

But it is not only research participants who face dangers. Researchers themselves have to confront health, wellbeing and safety challenges, and even traumatisation. They have to juggle ethics governance problems, either when there are no RECs that can guide their research or when existing RECs have insufficient insights into research in fragile settings. Practical problems, such as obtaining research ethics approval at short notice and informed consent in devastated locations, add to the burdens. Likewise, conflicts of interest become more problematic with research in fragile settings, because politics is never far from situations of peace and conflict or disaster.

The fact that all the burdens endured by research participants and researchers might ultimately still not pay off if the collected data is unreliable is a particularly daunting challenge.

As this review has demonstrated, research has the potential to worsen the situation of those already badly off, but the benefits of research to those communities can also be considerable.

When research in fragile settings is done well, it can reduce future suffering significantly. Kohrt et al. (2019) even argue that health research in humanitarian crises is an urgent global imperative.

> Globally, humanitarian crises – such as armed conflict, forced displacement, natural disasters and major disease outbreaks – affect more people today than at any point in recorded history. These crises have immense acute and long-term health impacts on hundreds of millions of people, predominantly in low and middle-income countries (LMIC), yet the evidence base that informs how humanitarian organisations respond to them is weak.

And indeed, to provide a better evidence base to inform humanitarian aid, research is urgently needed in areas such as health (ibid), but also areas such as climate change (Adaptation Fund 2024) or rebuilding countries post-war (Shanks and Paulson 2022).

The findings from this review of real-world challenges were needed to inform an ethics code that can then guide those researchers who understand the challenges confronting themselves and others when they work in fragile settings.

4.4 Potential Shortcomings of the Review and Their Mitigation

For this review two potential shortcomings are obvious. First, only one database was searched (Google Scholar). Second, only one primary researcher (DS) undertook the review, with checks by a second (LT), rather than two researchers working independently.

These two shortcomings were mitigated in the following way.

1. Three different strands of research worked on the same research questions, one review in English, one review in German and one survey.
2. The results from all three fed into the TRUST Code Supplement, which was validated through broad consultations.
3. Ethics codes can be refined and updated over time. Hence, should new findings emerge in future, it will be possible to act upon them.

Overall, one should bear in mind that ethics codes are almost never based on extensive research about real world ethics and integrity challenges. They are usually based on the drafters' views, with (sometimes broad) consultation. Hence, while there might be shortcomings, academically speaking, in a single-database review with only one primary researcher, the overall mission of providing a factual grounding for an ethics code was accomplished successfully, in comparison with most other ethics codes, exceptions being the TRUST Code (Schroeder et al. 2019) and the PREPARED Code (Chatfield and Singh 2025).

4.5 Conclusion

Research must take place in fragile settings in order to improve the situation in environments that are hostile to all those present. Such research faces major research ethics and integrity challenges. As illustrated in Chap. 2, it is important that any ethics code for fragile settings take account of real-world challenges and not be drafted remotely by faraway "experts".

Emulating the methodology of the TRUST Code (2018) (Schroeder et al. 2019), research ethics and integrity challenges for the TRUST Code Supplement (see Chap. 6) were identified through reviews, a survey and broad consultations.

This chapter has described a scoping review–informed evidence mapping of German sources. The review identified 16 major challenges: special burdens research participants face before the research; health, wellbeing and safety concerns for research participants; negative impacts arising from disingenuous or patronising research; the potential for direct harm caused by researchers; unrealistic expectations of research participants; retraumatisation risks for research participants; the protection-inclusion dilemma; the humanitarian misconception; payment-related challenges; misuse of research data and findings; health, wellbeing and safety concerns for researchers and support teams; ethics governance challenges; time-related pressures; conflicts of interest; challenges for the informed consent process; and issues related to data reliability.

These challenges were incorporated into the first draft of the TRUST Code Supplement, which was circulated for broad consultation.

References

Abdelkader S, Bach M, Narawitz L et al (2021) Forschungsethik als politische Verantwortung: Reflexionen zum Veröffentlichen in der Flucht*Migrationsforschung. In: Bach M, Narawitz L, Schroeder J et al (eds) FluchtMigrationsForschung im Widerstreit. Waxmann, Münster, pp. 211–225. https://www.waxmann.com/buecher/FluchtMigrationsForschung-im-Widerstreit. Accessed 29 Dec 2025

Adaptation Fund (2024) Addressing climate change adaptation in fragile settings and conflict-affected countries: lessons learned from the Adaptation Fund's Portfolio. https://www.adaptation-fund.org/wp-content/uploads/2024/02/climate-change-adaptation-in-fragile-settings-02.15.2024.pdf. Accessed 29 Dec 2025

Arksey H, O'Malley L (2005) Scoping studies: towards a methodological framework. Int J Soc Res Methodol 8(1):19–32. https://doi.org/10.1080/1364557032000119616

Billones R, Liwang JK, Butler K et al (2021) Dissecting the fatigue experience: a scoping review of fatigue definitions, dimensions, and measures in non-oncologic medical conditions. Brain Behav Immun–Health 5:100266. https://doi.org/10.1016/j.bbih.2021.100266

Buckley-Zistel S (2012) 'Ich bin dann mal weg.' Feldforschung im Studium der Friedens- und Konfliktforschung. Zeitschrift für Friedens- und Konfliktforschung 1(2):315–323. https://www.jstor.org/stable/48519338. Accessed 29 Dec 2025

Chatfield K, Singh M (eds) (2025) Research ethics and integrity during pandemics: developing the PREPARED code. SpringerBriefs in research and innovation governance. Springer, Cham, Switzerland. https://doi.org/10.1007/978-3-031-91324-2

Dittmer C, Lorenz DF (2018) Forschen im Kontext von Vulnerabilität und extremem Leid: ethische Fragen der sozialwissenschaftlichen Katastrophenforschung. Forum Qualitative Sozialforschung/Forum: Qualitative Social Research 19(3). https://doi.org/10.17169/fqs-19.3.3116

Erdmann N, Bartmann S, Haefker M et al (2022) Epilog: Erkenntnisse und Perspektiven für gesellschaftlich wahrgenommene Ausnahmesituationen. In: Bartmann S, Erdmann N, Haefker M et al (eds) Verstehendes Forschen in der Pandemie und anderen Ausnahmesituationen. Verlag Barbara Budrich, pp 159–168. https://doi.org/10.2307/j.ctv2tjdh29.11

Exo M (2017) Das übergangene Wissen: eine dekoloniale Kritik des liberalen Peacebuilding durch basispolitische Organisationen in Afghanistan. Transcript Verlag, Bielefeld. https://www.transcript-verlag.de/978-3-8376-3872-1/das-uebergangene-wissen/. Accessed 29 Dec 2025

Hearn J, Ssinabulya I, Schwartz JI et al (2019) Self-management of non-communicable diseases in low- and middle-income countries: a scoping review. PLoS ONE 14(7):e0219141. https://doi.org/10.1371/journal.pone.0219141

Hopia H, Heikkilä J (2019) Nursing research priorities based on CINAHL database: a scoping review. Nurs Open 7(2):483–494. https://doi.org/10.1002/nop2.428

Kabranian-Melkonian S (2015) Ethical concerns with refugee research. J Hum Behav Soc Environ 25(7):714–722. https://doi.org/10.1080/10911359.2015.1032634

Kohrt BA, Mistry AS, Anand N et al (2019) Health research in humanitarian crises: an urgent global imperative. BMJ Glob Health 4:e001870. https://doi.org/10.1136/bmjgh-2019-001870

Krause U (2016) Ethische Überlegungen zur Feldforschung: Impulse für die Untersuchung konfliktbedingter Flucht. CSS Working Paper No 20. Zentrum für Konfliktforschung der Philipps-Universität Marburg. https://doi.org/10.17192/es2019.0015

Krause U, Williams T (2021) Flexible Ethikgremien: Impulse zur Institutionalisierung ethisch verantwortlicher Feldforschung in der Konflikt- und Fluchtforschung. Soziale Probleme 32:97–113. https://doi.org/10.1007/s41059-020-00072-z

Lidz CW, Appelbaum PS, Grisso T, Renaud M (2004) Therapeutic misconception and the appreciation of risks in clinical trials. Soc Sci Med 58(9):1689–1697. https://doi.org/10.1016/S0277-9536(03)00338-1

Mak S, Thomas A (2022) Steps for conducting a scoping review. J Grad Med Educ 14(5):565–567. https://doi.org/10.4300/JGME-D-22-00621.1

Munn Z, Peters MDJ, Stern C et al (2018) Systematic review or scoping review? Guidance for authors when choosing between a systematic or scoping review approach. BMC Med Res Methodol 18:143. https://doi.org/10.1186/s12874-018-0611-x

Nature (2023) Scientific publishing has a language problem. Nat Hum Behav 7:1019–1020. https://doi.org/10.1038/s41562-023-01679-6

Niemann H, Schröder U (2020) Ein Feld in Bewegung: Friedens- und Konfliktforschung in herausfordernden Zeiten. Zeitschrift für Friedens- und Konfliktforschung 9:131–142. https://doi.org/10.1007/s42597-020-00033-1

OECD (2022) How fragile contexts affect the well-being and potential of women and girls. OECD Development Co-operation Directorate. OECD Publishing, Paris. https://www.oecd.org/en/publications/how-fragile-contexts-affect-the-well-being-and-potential-of-women-and-girls_c8b2287c-en.html. Accessed 29 Dec 2025

Peters MDJ, Marnie C, Colquhoun H et al (2021) Scoping reviews: reinforcing and advancing the methodology and application. Syst Rev 10(1):263. https://doi.org/10.1186/s13643-021-01821-3

Peters MDJ, Marnie C, Tricco AC et al (2020) Updated methodological guidance for the conduct of scoping reviews. JBI Evid Synth 18(10):2119–2126. https://doi.org/10.11124/JBIES-20-00167

Pfänder C (2021) Is German disappearing as a language of science? 30 November. Alumniportal Deutschland. https://www.alumniportal-deutschland.org/en/magazine/science-research/german-as-a-language-of-science/. Accessed 29 Dec 2025

Phillips T (2023) Coercion, exploitation, and undue inducement. In: Iltis AS, MacKay D (eds) The Oxford handbook of research ethics. Oxford University Press, New York, pp 154–172. https://doi.org/10.1093/oxfordhb/9780190947750.013.15

Pinillos-Franco S, Kawachi I (2022) Hostile attitudes toward immigrants and refugees are associated with poor self-rated health: analysis of 21 European countries. Soc Sci Med 301:114969. https://doi.org/10.1016/j.socscimed.2022.114969

Porzucki N (2014) Nobel Prize: how English beat German as language of science. BBC News, 12 October. https://www.bbc.com/news/magazine-29543708. Accessed 29 Dec 2025

PREPARED (n.d.) Our team. https://prepared-project.eu/our-team/. Accessed 29 Dec 2025

Ramos M (2025) Fragile contexts: the world's most dangerous places. 18 August. World Vision Canada. https://www.worldvision.ca/en/stories/what-are-fragile-contexts. Accessed 29 Dec 2025

Rice DB, Kloda LA, Levis B et al (2016) Are MEDLINE searches sufficient for systematic reviews and meta-analyses of the diagnostic accuracy of depression screening tools? A review of meta-analyses. J Psychosom Res 87:7–13. https://doi.org/10.1016/j.jpsychores.2016.06.002

Schrader H (2024) Ostwärts! Lehrforschungen der Jahre 1998–2023 in fremdkulturellen Kontexten. DSS Working Paper Nr. 83. Institut für Gesellschaftswissenschaften, Otto-von-Guericke-Universität Magdeburg. https://www.soz.ovgu.de/isoz_media/downloads/arbeitsberichte/83_Schrader.pdf. Accessed 29 Dec 2025

Schroeder D, Chatfield K, Chennells R et al (2024) Vulnerability revisited: leaving no one behind in research. Springer Nature Switzerland AG, Cham, Switzerland. https://doi.org/10.1007/978-3-031-57896-0

Schroeder D, Chatfield K, Singh M et al. (2019) Equitable research partnerships: a global code of conduct to counter ethics dumping. Springer Nature, Cham, Switzerland. https://doi.org/10.1007/978-3-030-15745-6

Schwenken H (2015) Rechtlos, aber nicht ohne Stimme: politische Mobilisierungen um irreguläre Migration in die Europäische Union. Transcript Verlag, Bielefeld. https://www.transcript-verlag.de/978-3-89942-516-1/rechtlos-aber-nicht-ohne-stimme/. Accessed 29 Dec 2025

Shanks K, Paulson J (2022) Ethical research landscapes in fragile and conflict-affected contexts: understanding the challenges. Res Ethics 18(3):169–192. https://doi.org/10.1177/17470161221094134

Sieber Egger A (2011) Krieg im Frieden: Frauen in Bosnien-Herzegowina und ihr Umgang mit der Vergangenheit. Transcript Verlag, Bielefeld. https://www.transcript-verlag.de/978-3-8376-1624-8/krieg-im-frieden/. Accessed 29 Dec 2025

van Enst WA, Scholten RJ, Whiting P et al (2014) Meta-epidemiologic analysis indicates that MEDLINE searches are sufficient for diagnostic test accuracy systematic reviews. J Clin Epidemiol 67(11):1192–1199. https://doi.org/10.1016/j.jclinepi.2014.05.008
Włostowski Ł (2024) Camp als Commons? Ein selbstverwaltetes Projekt von Geflüchteten in Lavrio (Griechenland). Alice Salomon Hochschule Berlin. https://doi.org/10.58123/aliceopen-604

Chapter 5
Ethics and Integrity Challenges in Fragile Settings: A Survey of Experienced Researchers

Abstract Conducting research in fragile settings can be fraught with challenges and risks for participants, researchers and communities alike, together with considerable feasibility constraints. Within such contexts, research ethics and research integrity concerns are often intensified, and novel challenges may arise as stakeholders operate in unstable, unsafe and unpredictable environments. This chapter outlines the empirical foundation underpinning the development of the TRUST Code Supplement—A Global Code of Conduct for Research in Fragile Settings. It draws upon a survey of researchers with first-hand experience in fragile contexts, whose insights served both to validate and expand upon challenges identified in the existing literature, while also checking for potential critical gaps. Thirty-six researchers completed the online questionnaire, which comprised both open and closed questions. Alongside comprehensive literature reviews and extensive consultation, the survey findings helped to establish an evidence-informed foundation for the TRUST Code Supplement, thereby enhancing its relevance and real-world applicability.

Keywords Research ethics · Research integrity · Research ethics in fragile settings · Research ethics in conflict-affected areas · Humanitarian research

5.1 Introduction

For ethics codes to be effective, they must address the challenges that researchers actually encounter. Accordingly, a risk-based approach was adopted for the development of the TRUST Code Supplement—A Global Code of Conduct for Research in Fragile Settings. The key strength of the risk-based approach is that it grounds the requirements appearing in the ethics code in real-world events, informed by a diversity of perspectives and experiences (Chatfield and Singh 2025). This enhances the practical value of the code and its relevance for the users. In order to uncover the actual risks and dilemmas that arise when conducting research, the approach requires considerable effort. For the TRUST Code Supplement this involved comprehensive

D. Schroeder et al., *Research Ethics and Integrity in Fragile Settings*, SpringerBriefs in Research and Innovation Governance,
https://doi.org/10.1007/978-3-032-36141-7_5

literature reviews (see Chaps. 3 and 4), empirical research and extensive consultations to capture the ethical risks and dilemmas that can arise when conducting research in fragile settings of different types. This chapter focuses specifically upon the empirical input, which took the form of a survey of researchers who had personal experience of working in fragile settings.

Research in fragile settings is characterised by intersecting challenges that shape both the feasibility of fieldwork and the ethical dilemmas that researchers face (Idris 2019; Shanks and Paulson 2022). Many of these challenges are context-specific, shaped by local cultural, political, economic and environmental factors (Woodward et al. 2017; Diaconu et al. 2020). A host of such challenges are well described in published literature (see Chaps. 3 and 4). Nevertheless, while literature reviews can provide valuable breadth and theoretical grounding, they are limited in their ability to capture the lived, context-specific realities of conducting research in fragile settings.

To address this gap, our empirical study sought primary data in the form of first-hand, lived accounts. First-hand accounts were critical for three reasons: first, to validate insights from the existing literature; second, to provide clarification and elaboration on key issues; and third, to expose potential gaps in knowledge that might not be evident in published work. It was essential to these ends for us to seek out the lived experiences of those who had personally navigated fragile-setting environments.

The specialist population we targeted comprised researchers who had direct experience of conducting studies in any type of fragile setting, across geographical regions. By engaging with researchers from varied contexts, the study sought to gain a nuanced understanding of research ethics and research integrity challenges as manifest in a range of settings and contexts. This is why the *Research Ethics and Integrity Challenges in Fragile Settings* questionnaire was designed and the survey implemented.

This chapter outlines the process of constructing the questionnaire and implementing the survey, including details about securing ethics approval, the recruitment, the data collection and findings, and the implications for the development of the research ethics and integrity code.

5.2 Constructing the Questionnaire

The overarching aim of the survey was to inform the development of a new code upholding research ethics and integrity in fragile settings. The specific objectives were as follows:

1. to validate insights derived from existing literature on research ethics and research integrity challenges in fragile settings
2. to provide further clarification and elaboration on key issues encountered in practice that relate to research ethics and integrity

3. to identify potential gaps in existing literature-based knowledge by bringing to light challenges that were not found in published sources.

To address these objectives, the survey questions were designed to align closely with the existing literature. This enabled corroboration of the challenges reported in published sources, and an assessment of whether they reflected those encountered in practice accurately, while it also allowed respondents to elaborate on the issues in greater depth. Hence, the majority of questions in the survey took the form of statements about documented challenges in research ethics and integrity in fragile settings. Respondents were asked to indicate whether or not they had experience of these challenges and to add qualitative comments about their experiences. For example:

Question 4 The health and safety of research participants is difficult to maintain in fragile settings, especially in conflict and post-conflict settings (Select one)

- Yes
- No
- I don't know

Please comment (open-ended, free text box).

The questionnaire comprised 40 questions, organised into the thematic areas set out in Table 5.1.

Questions 1–3 were designed to collect contextual information, while the last question (40) provided an open-ended opportunity for participants to share experiences beyond those identified in the survey. This question asked:

Table 5.1 Topic groupings used in the survey

Question numbers	Topics
1–3	Establishing the types of fragile settings in which the survey respondents had conducted research; positionality in relation to the setting; and sources of ethics approval for their research
4–8	Impacts upon the research participants
9–14	Impacts upon the local communities
15–25	Impacts upon the research
26–32	Impacts upon the researcher
33–36	Misperception challenges
37–39	Access to the research participants
40	Open question

> Have you experienced—directly or indirectly—any other ethics and integrity challenges during your research in fragile settings, which are not included above?
> If so, please describe them briefly (open-ended, free text box).

The responses to this question facilitated the identification of additional challenges not found in the literature.

Thus, the questionnaire was designed to capture both qualitative and quantitative data through the use of open and closed questions. The closed questions generated quantitative data that helped to establish context and quantify the extent to which respondents had experienced (either directly or indirectly) the identified risks and challenges. The open questions generated qualitative data, as unlimited space was provided for respondents to share their personal experiences, observations and opinions.

The questionnaire was drafted by Doris Schroeder, Kate Chatfield (KC) and Hazel Partington (HP). To test clarity and relevance, the survey was piloted with four researchers who had direct experience of working in fragile contexts. Their feedback led to several further adjustments. Among the responses, one participant provided particularly encouraging feedback:

> It certainly prompted me to reflect upon many issues I have encountered during my field research … I felt it was a truly worthwhile experience to go over these questions.

The questionnaire was developed on Jisc Online Surveys (https://onlinesurveys.ac.uk/) (formerly Bristol Online Surveys) to enable global reach via an online survey.

5.3 Ethical Considerations

Fragile settings are often characterised by major intersectional challenges such as conflicts, instability and limited institutional support. Research in these settings is therefore inherently sensitive, raising complex concerns about power dynamics, participant safety, researcher safety and so on. To encourage open responses, participation in the survey was fully anonymous and respondents were also asked to refrain from including any information that could identify themselves or others. Any identifiable information that was provided was permanently removed by the study team before analysis.

Ethics approval for the study was obtained from the University of Lancashire's Health Ethics Panel (Health 01174). Participant consent was secured via an information page and consent form on the first page of the online survey. The survey and participant information were designed to be GDPR compliant.

5.4 Recruitment

Recruitment followed a snowballing strategy (Tenzek 2017), beginning with members of the PREPARED consortium, several of whom had prior experience in fragile settings. Members of the consortium shared the survey invitation through their professional networks, which included universities, healthcare institutions, nongovernmental organisations (NGOs), Médecins Sans Frontières and the Drugs for Neglected Diseases initiative (DNDi). In addition, the invitation to participate and the survey link were publicly available on the PREPARED consortium website.

The survey was open to individuals aged 18 and over who had conducted any form of research in any type of fragile setting other than pandemic contexts. The term “fragile settings” was defined for potential respondents as.

> environments such as conflict or post-conflict zones, regions which recently experienced natural disasters; refugee camps under serious resource constraints; or high-crime settings such as high-crime informal settlements.

The information stated explicitly that we were not referring to pandemic settings, which are addressed specifically in the PREPARED Code (2025). Eligibility to participate was determined via self-identification at the start of the survey. The survey remained open for 4 months, from November 2024 until February 2025.

5.5 Analysis

Quantitative analysis of the closed questions was limited to descriptive statistics (Kaur et al. 2018) undertaken via the online survey’s analysis functions before being downloaded to secure university storage. Of primary interest was the level of agreement or disagreement among respondents regarding their experiences of the described challenges, as a means of assessing the authenticity of challenges previously identified in the literature.

Qualitative data were downloaded, anonymised (if necessary) and analysed by KC and HP. There are many ways of analysing qualitative data (Schwandt 2014); different analytic approaches serve different purposes, highlight different aspects of data, and generate distinct kinds of findings. Most vital is to align the chosen method of analysis closely with the research aims and objectives (Jansen et al. 2010; Wright et al. 2016). Accordingly, the analysis in this study focused on identifying and examining information directly relevant to the development of a research ethics and integrity code. This process involved assessing whether the range of challenges identified in the literature corresponded with practice, as well as exploring how these challenges are encountered in real-world contexts. As a first step, potential exacerbating factors and mitigating measures were deduced, where possible, from the free-text responses. These factors aid understanding of what might lead to ethics and integrity issues, as well as what might be needed to address them.

5.6 Results

Thirty-six researchers completed the survey. The results below are organised by the thematic areas that appeared in the questionnaire (see Table 5.1).

5.6.1 Settings Covered

Survey respondents were presented with a list of types of fragile settings and asked to select those in which they had conducted research. Of the 36 respondents, 21 (60%) had worked in conflict settings, 10 (29%) in post-conflict settings, 7 (20%) in high-crime contexts, 5 (14%) in areas affected by natural disasters, 9 (26%) in refugee camps under severe constraints and 4 (11%) in the aftermath of riots or other violent events (Fig. 5.1).

Respondents who selected "Other" were invited to provide further details. Seven respondents (11%) identified additional fragile contexts, including research with leprosy patients, migrant populations, victims of gender-based or sexual violence and remote Indigenous populations; research in regions experiencing drought; and research during the demolition of informal housing settlements.

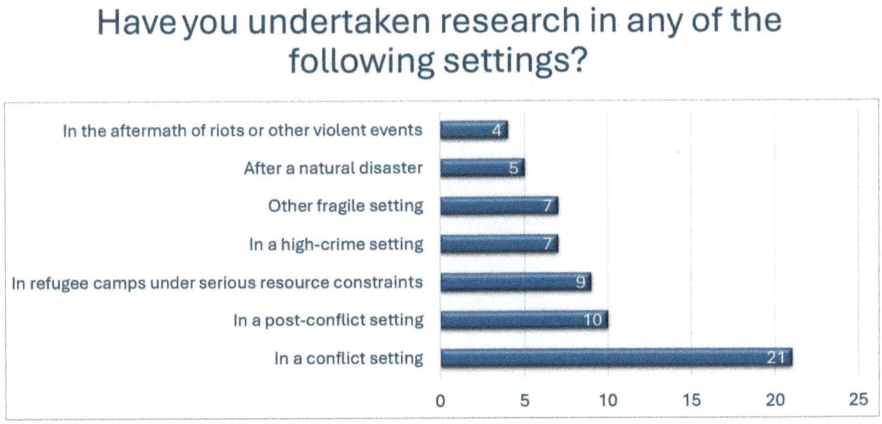

Fig. 5.1 Types of settings encountered

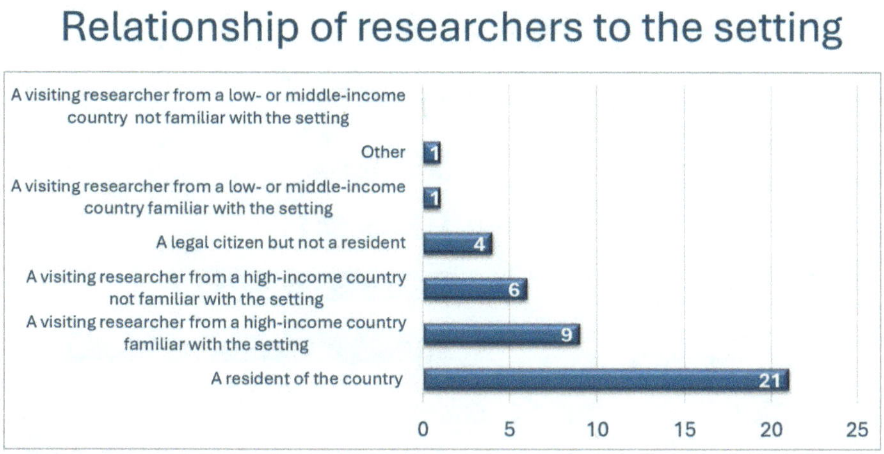

Fig. 5.2 Position of researcher in relation to settings

5.6.2 Positionality in Relation to the Research Setting

Of the 36 respondents, 21 (60%) were residents of the country where the research was conducted, 9 (26%) were visiting researchers from high-income countries (HICs) who were familiar with the setting, 6 (17%) were visiting researchers from HICs who were unfamiliar with the setting, 4 (11%) were legal citizens but not residents of the country in which the research took place, and 1 (3%) was a visiting researcher from a low-income country (LMIC) who was familiar with the setting (Fig. 5.2).

The respondent who selected "Other" had worked as a humanitarian aid worker (medical coordinator) for an organisation involved in a therapeutic trial.

5.6.3 Research Ethics Approval

Survey respondents were asked to identify the body from which research ethics approval for their studies had been obtained. Of the 36 respondents, 23 (66%) reported approval from their own institution's research ethics authority, while 14 (40%) had obtained approval from relevant authorities in the country where the research took place, 9 (26%) from a local community body and 2 (6%) from the funder's research ethics authority (Fig. 5.3).

Additional responses in the "Other" category included approval from NGOs or similar organisations and from the relevant Ministry of Health. Two respondents (6%) indicated that their studies had not received research ethics approval. Some respondents may have received approval from more than one source.

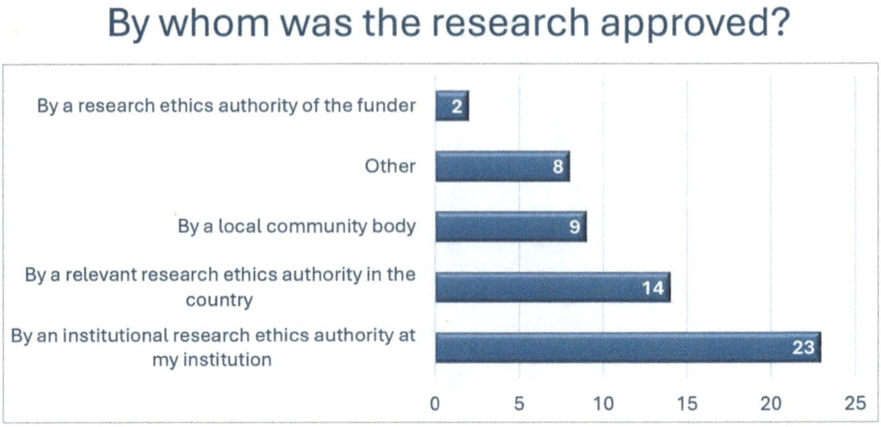

Fig. 5.3 Body approving research

For the remainder of the questionnaire, respondents were given the following instructions:

> The remaining questions in this survey concern documented research ethics and integrity challenges for research in fragile settings. Please tick yes if you have experience of the following documented challenges and use the comments boxes if you wish to tell us something about the experience.

Accordingly, respondents who answered "Yes" can be understood to have had personal experience of, or to be familiar with, the challenges described.

These results are summarised below by thematic area and ordered by level of agreement. The tabulated results also show related exacerbating factors and mitigating measures that could be inferred from the free-text responses. Blank cells in the tables indicate where no inferences could be drawn from the data.

5.6.4 *Impacts on Research Participants*

While there was consistent agreement that maintaining the health and safety of participants can be difficult in fragile settings, there was less certainty about whether participants in fragile settings are more open to undue inducement. Caution was advised regarding assumptions about participants' motives: while some may indeed be vulnerable to undue inducement, the potential risks of being a research participant make others more reluctant to participate, regardless of their circumstances (Table 5.2).

Table 5.2 Impacts on research participants

Challenge	Agreement level	Exacerbating factors	Mitigation measures
The health and safety of research participants is difficult to maintain in fragile settings, especially in conflict and post-conflict settings	Yes = 86% No = 8% Don't know = 6%	• Increased vulnerability of participants, e.g., through poverty, trauma, mental health issues, lack of power • Unhealthy and dangerous environment • Disrupted healthcare services • Mobilisation challenges for participants and healthcare personnel • Digital and/or paper trails that might render participants identifiable. • Risk of dual use, i.e., use of sensitive information for harmful purposes	• Take precautions to avoid compromising participant safety, e.g., careful risk assessment, secure data collection methods, flexible scheduling, safe meeting locations; have emergency response protocols in place; work with NGOs; work closely with trusted locals; only work through established contacts • Adapt methods to enhance safety, e.g., interact by phone rather than in person • Transfer clinical trial patients to safer sites
Involvement in research can lead to re-traumatization for research participants	Yes = 72% No = 17% Don't know = 11%	• Special sensitivity of some projects (e.g., investigating the experiences of refugees who have experienced traumas)	• Require trauma-informed practices: careful design of questions, training of researchers and relevant support, like counselling services post-interview
Ensuring confidentiality and/or privacy is significantly challenged in fragile settings, especially if the area is controlled by armed actors	Yes = 69% No = 17% Don't know = 14%	• Relative conspicuousness of foreign researchers • Problems potentially created by presence of researchers • Armed actors needing to approve the research • Armed actors demanding information • Higher risk of data breaches • Greater challenges in small communities where reidentification is easier • Relative conspicuousness of focus groups	• Don't be naive • Remove all identifying factors

(continued)

Table 5.2 (continued)

Challenge	Agreement level	Exacerbating factors	Mitigation measures
Informed consent processes might be very problematic	Yes = 69% No = 28% Don't know = 3%	• Perceived risk of signing a document • Great need making people more vulnerable to coercion (e.g. willing to participate just to get a free meal) • Greater frequency of logistical constraints, such as transport difficulties • Relatively high suspicion/caution • Living with fear/anxiety/trauma hampering people's ability to take in lots of information • Impracticability of lengthy procedures	• Use researchers from the community to help with seeking consent
Due to the extremity of their situation and potential desperateness, research participants are more likely to be open to undue inducements than in other settings	Yes = 61% No = 14% Don't know = 25%	• Situation of abject poverty • Lack of other access to support for needs (food, healthcare etc.) • Willingness to do anything to improve one's situation	Manage needs of participants carefully

Several respondents noted that the very nature of fragile settings can make it difficult to guarantee participants' confidentiality and privacy. Beyond standard data security concerns, the conspicuousness of foreign researchers makes it easier for others to identify those interacting with them. In addition, the presence of armed actors can heighten risks for participants—for example, if such actors demand information about those taking part in the research.

Although many challenges were reported in obtaining informed consent, a notable proportion of respondents (28%) indicated that they had not experienced any difficulties. However, several emphasised the importance of maintaining flexibility in the consent process.

5.6.5 Impacts on Local Communities

Cultural and communication challenges in fragile settings emerged as the most frequently experienced challenge for communities (72% of respondents). While comments indicated that such challenges are not confined to times of crisis, they may be heightened in fragile contexts depending on circumstances. A majority of respondents (64%) also acknowledged the difficulty of returning findings to local populations in fragile settings. However, some emphasised that concerted efforts should be made to do so wherever possible, framing this as an ethical responsibility. Similarly, just over half of the respondents reported experiencing challenges with community engagement, yet their comments maintained that engagement should still be pursued wherever feasible (Table 5.3).

Fewer than half of respondents reported challenges with fair remuneration, the protection of resources or humanitarian efforts, and the limited number of comments in these areas precluded firm inferences from the data.

5.6.6 Impacts on Research

Eleven potential impacts upon research were presented in the questionnaire, and at least half the respondents had experienced ten of these. This demonstrates clearly the breadth of challenges to the quality of research in fragile settings. The most commonly experienced challenge (72%), that of inconsistent and unsystematic data collection, is perhaps not surprising given the inherent unpredictability of fragile settings. Other challenges might be less obvious to a foreign researcher: for instance, that research in fragile settings can appear to follow colonialist practices. In this regard, respondents pointed to the fragility of the setting exacerbating power imbalances, and to ethics dumping as the "biggest elephant in the room" (Table 5.4).

Table 5.3 Impacts on local communities

Challenge	Agreement level	Exacerbating factors	Mitigation measures
Cultural and communication challenges are significantly heightened in fragile settings	Yes = 72% No = 17% Don't know = 11%	• Suspicion towards outsiders heightened by crisis situation • Issues/barriers caused when local researchers are affiliated to certain cultural/religious groups	• Be as respectful as possible • Embed the research within humanitarian projects
It can be very difficult to return findings to local populations	Yes = 64% No = 28% Don't know = 8%	• Resettlement/migration/displacement of participants • Politically sensitive findings • Safety issues for researchers	• Put findings on a website • Return findings to the community if not the exact participants
There is often insufficient time to develop meaningful community engagement strategies, for instance, after natural disasters	Yes = 54% No = 29% Don't know = 17%	• Unavailability of usual meeting places and platforms • Time constraints • People being on the move	• Be responsive and adapt quickly when making plans • As a first response, set up social systems and structures that can be utilised • If the research plans a solution, the community will agree to engage
It can be difficult to obtain fair remuneration for local support (e.g. interpretation) from funders	Yes = 43% No = 34% Don't know = 23%		
Researchers' presence may divert limited resources (e.g. food, health care, sleeping opportunities) from local populations	Yes = 31% No = 36% Don't know = 33%		
Research can impede, delay or weaken humanitarian efforts	Yes = 26% No = 46% Don't know = 29%		

Blank cells in the table indicate where no inferences could be drawn from the data

Table 5.4 Impacts on research

Challenge	Agreement level	Exacerbating factors	Mitigation measures
Data collection might be inconsistent and unsystematic	Yes = 72% No = 17% Don't know = 11%	• Movement of participants • Higher degree of uncertainty • Greater caution among participants	• Revise the sampling strategy • Adjust the sample size • Change the data collection methods • Move the study to a different area
There is a risk of duplicating research due to lack of coordination	Yes = 69% No = 19% Don't know = 11%	• Different agencies working in the locale • Some researchers not caring what others are doing • Research often conducted in a hurry	• Aim for better communication between research teams
The thematic focus of research is often dominated by topics that researchers from other countries can obtain funding for	Yes = 69% No = 14% Don't know = 17%	• The influence of the priorities of funding bodies. • Greater media and academic attention for certain themes	
Gold standard methods of research are not always possible in fragile settings, e.g. random sampling	Yes = 61% No = 25% Don't know = 14%	• Chaotic situations • Lack of complete population lists (making random sampling impossible) • Impossibility of implementing the usual methods	• Determine an alternative "gold standard" for the setting
It is often difficult to get local ethics approval in fragile settings	Yes = 58% No = 31% Don't know = 11%	• Dismantled state of social structures • Effect of fragile setting on REC members • Suspension of some RECs	• Seek approval/permission from another local authority • Plan well in advance of field work

(continued)

Table 5.4 (continued)

Challenge	Agreement level	Exacerbating factors	Mitigation measures
Research in fragile settings can appear to follow colonialist practices with little involvement of local researchers	Yes = 58% No = 25% Don't know = 17%	• Power imbalances exacerbated by fragility • Research leadership often by outsiders, as locals are preoccupied by immediate demands of the crisis • Less time for community engagement	• Adapt research approaches to take account of local preferences and situations
There is a risk of focusing on emergency/acute care/infectious diseases to the detriment of non-communicable diseases that can be the main driver for morbidity and mortality in fragile settings	Yes = 56% No = 8% Don't know = 36%	• Health systems in fragile settings opting to prioritise dealing with infectious diseases due to their immediate, visible impacts • Research direction determined by funding • Very reactive responses	
It can be very difficult to validate findings with local populations	Yes = 56% No = 28% Don't know = 17%	• Lack of correct and consistent information • Lack of baseline data • Displaced populations • Re-entry difficulties for researchers • People being unable to remember.	• Plan for this • Get the local population "on board" • Validate with the community rather than the participants
It can be difficult to recruit local partners in fragile settings	Yes = 56% No = 36% Don't know = 8%	• People being on the move or otherwise engaged • Lack of support or funding for the research • Fear and mistrust of researchers • Lack of skills and training	• Put effort into awareness raising

(continued)

Table 5.4 (continued)

Challenge	Agreement level	Exacerbating factors	Mitigation measures
There is a risk of an uncoordinated influx of researchers from outside of the country in which the research is taking place	Yes = 50% No = 28% Don't know = 22%		
Due to the need for speed, scientific rigour might be diminished in study design	Yes = 43% No = 40% Don't know = 17%	• Volatility and displacement challenges	• Be realistic about what can be achieved and honest about limitations • Insist on pre-planning • Adapt data collection methods • Ensure that methods are appropriate to the context

Blank cells in the table indicate where no inferences could be drawn from the data

Twenty-one respondents (58%) had experienced difficulties in attaining local ethics approval in fragile settings. Several pointed out that local ethics committees are themselves likely to be impacted by the fragility of their setting. Nevertheless, the need to seek local ethics approval, wherever possible, was emphasised, along with an appeal to researchers not to use a fragile setting as an excuse not to seek ethics approval.

Unsurprisingly, planning, flexibility in design and the provision of adequate funding were often cited as potential mitigating factors for the issues identified.

5.6.7 Impacts on Researchers

The first three challenges listed in Table 5.5 were also the most frequently reported across the entire questionnaire, experienced by 94%, 91% and 89% of respondents, respectively. From these findings we can assume that the vast majority of individuals who conduct research in fragile settings experience stress, as well as health and safety concerns.

Reported sources of stress ranged from physical hardships to psychological strains, such as witnessing or hearing about distressing experiences, facing threats to life, and experiencing moral distress from being unable to help everyone. Additionally, it was noted that risks may extend beyond the data collection phase. For example, there may be repercussions for the safety of researchers and their families from the published findings. Moreover, no respondents provided comments that could be interpreted as mitigation strategies for stress. Consequently, despite this being the most frequently reported challenge, no corresponding mitigation measures are included in Table 5.5.

The health and safety of researchers, in particular field data collectors, can be difficult to maintain in fragile settings with numerous reported threats. Comments told stories of researchers being kidnapped by an armed group, and finding themselves stranded in high-crime settings or in close proximity to violence, and also of exposure to other logistical, physical and psychological dangers. Additionally, 21 respondents (58%) agreed that researchers may face significant local hostility. Some respondents differentiated between local community hostility and hostility from the government, local authorities, decision-makers or groups engaged in armed conflict. However, it was also pointed out that experiences of local hostility are not unique to fragile settings.

5.6.8 Misperception Challenges

Most respondents (83%) agreed that researchers can raise expectations about benefits on which they cannot deliver. Comments pointed to the need for researcher transparency, because it is not unusual for people to request assistance, and it is

Table 5.5 Impacts on researchers

Challenge	Agreement level	Exacerbating factors	Mitigation measures
Researchers can experience significant stress themselves in fragile settings	Yes = 94% No = 3% Don't know = 3%	• Poverty-stricken surroundings • Having to work remotely • Suspicion and lack of trust of researchers • Demands from the community • Not being able to help	
The health and safety of researchers, in particular field data collectors, can be difficult to maintain in fragile settings	Yes = 91% No = 6% Don't know = 3%	• Conflict • Instability • Poor infrastructures	• Make plans • Involve the local team in risk management and mitigation • Delegate to a local lead who can assess the situation • Take or avoid certain forms of transportation, curfews etc. • Engage ongoing and impartial security oversight to protect the in-field researchers
Researchers can find it difficult to deal with traumatized people or with people in a setting of fear	Yes = 89% No = 6% Don't know = 6%	• Normal systems not working • Trust issues	• Anticipate and plan accordingly (including identification of support systems and resources) • Give researchers appropriate training in trauma-informed responses and psychological support • Include a trauma expert as part of the team • Include oversight by mental health practitioners

(continued)

Table 5.5 (continued)

Challenge	Agreement level	Exacerbating factors	Mitigation measures
It can be unclear what researchers should do when they encounter human rights abuses	Yes = 81% No = 14% Don't know = 6%	• Access to participants and communities gained via the perpetrators • Disorderly environments • Lack of safety when approaching police • Different interpretations of human rights • Cultural complexity	• Anticipate and plan, including establishing protocols
Researchers may feel under pressure to ignore risks to themselves when they want to continue their research, e.g. to obtain further research data	Yes = 78% No = 8% Don't know = 14%	• Locally recruited researchers entering the field • Precarious employment status	
It can be very difficult to obtain permission to travel to a fragile setting	Yes = 78% No = 8% Don't know = 14%	• Health and safety policies • Risk-averse institutions/ ethics committees • Insurance restrictions	• Consult the in-country travel advice
Researchers may face significant local hostility (62)	Yes = 58% No = 11% Don't know = 31%	• High levels of fear and mistrust • Conflict and post-conflict zones	• Plan ahead and do the groundwork • Involve locals

Blank cells in the table indicate where no inferences could be drawn from the data

Table 5.6 Misperception challenges

Challenge	Agreement level	Exacerbating factors	Mitigation measures
Researchers can raise expectations about benefits they cannot deliver on	Yes = 83% No = 6% Don't know = 11%	• The research being focused on areas of dire need for the population	• Be transparent about the actual potential impacts • Communicate clearly
People confuse research with humanitarian help when researchers are also part of humanitarian efforts (i.e. when they have dual roles)	Yes = 72% No =3% Don't know = 25%	• Researchers who are also health providers	• Obtain appropriate informed consent
Researchers who are not part of humanitarian efforts can be wrongly perceived as helpers	Yes = 69% No = 3% Don't know = 28%	• People who are desperate and easily confused • People being keen to ask for help	• Communicate clearly • Take time to explain the research well
There is an expectation that researchers will provide humanitarian support (e.g. use their car to drive injured to field hospitals)	Yes = 67% No = 3% Don't know = 31%	• The possibility that participants might be injured or attacked	• Plan for this, as it is a reasonable expectation • Discuss with local colleagues how to respond

understandable that populations in fragile settings may see researchers as potential helpers (Table 5.6).

5.6.9 Access to Research Participants

Most respondents (61%) agreed that researchers can tend to use the same gatekeepers to access the same research populations, but this was noted as a general problem in foreign settings and not specific to fragile settings. The same might also be said for the other two challenges listed in Table 5.7, although one could assume that bribery is more likely to happen when people are in desperate and/or ungoverned situations. However, the responses were not clear about how or whether this actually happens.

Table 5.7 Access to research participants

Challenge	Agreement level	Exacerbating factors	Mitigation measures
Researchers can tend to use the same "gatekeepers" to access the same research populations, thereby creating over-research in one group and under-research in others	Yes = 61% No = 6% Don't know = 33%	• No time to build trust	
The need for "gatekeepers" might mean that researchers only see what they are meant to see	Yes = 58% No = 3% Don't know = 39%	• Gatekeepers with political backing	• Cooperate with local institutions
It can be possible to gain access to research populations by providing bribes to "gatekeepers"	Yes = 49% No = 11% Don't know = 40%	• Lack of time to build new relationships	• Anticipate this and plan for it

The blank cell in the table indicates where no inferences could be drawn from the data

5.6.10 Other Ethics and Integrity Challenges

Respondents were asked:

> Have you experienced—directly or indirectly—any other ethics and integrity challenges during your research in fragile settings, which are not included above? If so, please describe them briefly.

15 respondents answered this question. Several responses mentioned issues that had already been addressed in comments on previous questions, for example, participants being reluctant to participate in research due to a fear of repercussions, the unfair treatment of local researchers including in terms of authorship and recognition, ethics dumping issues, participants' expectations of aid from the researcher, the neglect of noncommunicable diseases, and risks involved in conflict situations with military occupation.

The novel issues mentioned included:

- problems with local and government authorities, such as
  - threats directed at researchers if they did not alter their findings to portray the authorities in a better light
  - pressure from government authorities to shape findings to support an official narrative
  - misuse of findings by authorities for political purposes,
- balancing the wishes of a study population against the need for impartiality in reporting

- the impacts of cultural and gender dynamics on female researchers
- whether findings could be published in peer-reviewed journals when local ethics approval had not been available
- frustration with the caution of research ethics committees in non-fragile settings about the conduct of research in high-risk settings.

5.7 Interpreting the Findings

Overall, our findings corroborated the challenges identified in the literature, although some appeared to be more generally experienced than others.

By far the greatest commonality concerned impacts upon the researchers themselves, who had experienced significant stress, difficulties maintaining their own health and safety, and the challenges of dealing with traumatised people or people living in fear. Further, most respondents had felt under pressure to continue their research and ignore risks to themselves, indicating that conducting research in a fragile setting entails not only methodological and ethical challenges but also a substantial personal cost.

There are also significant potential costs for research participants; of the 36 respondents, 31 had experienced difficulties in ensuring the health and safety of participants, including concerns about retraumatisation and the preservation of confidentiality and privacy. Informed consent processes were seen to be problematic in situations where there could be lack of trust and (justified) fears about signing documents. The consent process can also entail added complexities. For instance, as many as 72% of the respondents had experience of people confusing research with humanitarian help when researchers were also involved in humanitarian efforts, and 61% had seen desperation render participants more open to undue inducements.

There was broad experience of heightened cultural and communication challenges (72%) and problems with returning findings to local communities (62%), but there was far less experience of other potential impacts on the local community, such as challenges to developing meaningful community engagement, obtaining fair remuneration for local support, or the presence of researchers diverting limited resources and thus impeding or weakening humanitarian efforts.

Our findings confirmed that conducting research in a fragile setting can have significant impacts upon the quality of the research. Most of our respondents had experienced inconsistent or unsystematic data collection (72%), duplication of research through lack of coordination (69%), the domination of research topics set by other countries (69%), and problems employing gold standard methods of research (61%). Given the risks to participants, communities and researchers, this raises serious concerns about the potential benefits and burdens of the research.

Unexpectedly, few additional challenges beyond those enumerated in the questionnaire were volunteered. The most notable comments from respondents concerned difficulties with local and government authorities, including the potential misuse of findings and the exertion of undue pressure. While these issues warrant consideration, the available data were insufficient to draw firm conclusions regarding their broader implications.

The open-ended comment boxes allowed respondents to elaborate on the challenges described. All of the questions generated comments, though their volume and detail varied, with less commonly encountered challenges attracting fewer and briefer responses. Nevertheless, the richness of the data offered deeper insight into how these challenges are experienced and the strategies employed to address them. Although we did not explicitly ask about exacerbating or mitigating factors, these could often be extracted or inferred from the responses. Such factors provide essential contextual information for anticipating, avoiding and addressing challenges in the field.

5.7.1 From Challenges to Risks to an Ethics Code

As mentioned earlier, this empirical study was undertaken as part of the risk-based approach to the development of the TRUST Code Supplement, an approach designed to ensure that the requirements appearing in the ethics code are grounded in real-world events and informed by diverse perspectives and experience (Chatfield and Singh 2025). Following validation and elucidation of the challenges, the next step in developing the code entailed translating the reported challenges into the potential risks for research ethics and research integrity in fragile settings that need to be addressed by the code. As this is the same approach as that discussed in two earlier SpringerBriefs, one dealing with the TRUST Code (Schroeder et al. 2019) and the other with the PREPARED Code (Chatfield and Singh 2025), it will not be set out in detail here.

5.7.2 Reliability of the Results

In this study, 36 people shared their experiences of the challenges they encountered when conducting research in fragile settings. While this number of respondents might appear small for survey research (Rahi et al. 2019), there are several reasons why we consider it sufficient to meet our stated objectives.

First, our objectives were primarily confirmatory and explanatory. Rather than seeking to address an open research question, we sought to validate and further elucidate research ethics and integrity challenges that had already been identified in the literature. In this regard, we could argue that even a single positive response about experience of a challenge might be taken as confirmation that it is encountered in real-world contexts. Further, while elucidation of the challenges would benefit from input from a range of perspectives, every type of fragile setting listed in the first question was represented by at least five respondents (Fig. 5.1), and five other types of settings were mentioned under the "Other" category.

Second, a significant criticism of research in fragile settings, especially in humanitarian settings, is that it draws few researchers from the countries affected by the crises concerned and is dominated by researchers from Western universities (Antouly et al. 2024). If so, this would mean that our findings could be susceptible to "foreign gaze", which describes findings influenced by entrenched power asymmetries in global health partnerships (Abimbola 2019) and/or cultural insensitivity (Chatfield et al. 2020). However, in our survey, a majority of respondents (60%) were residents of the country where their research was undertaken, and only 26% had conducted research as outsiders from high-income countries in low- or middle-income settings.

Third, the total pool of researchers from which we were recruiting (those who had first-hand experience of working in a fragile setting) may appear to be quite small. A systematic review of public health interventions in humanitarian contexts revealed a scarcity of research and major gaps in the research base, with only 345 eligible studies conducted over more than 30 years (Blanchet et al. 2017). Conducting research in fragile settings is difficult and funding is scarce (Antouly et al. 2024). While there are indications that research has increased in recent years, it remains insufficient (Leresche et al. 2020), and typically in crisis-affected settings there will only be a small number of trained researchers (Idris 2019). Against this backdrop, the recruitment of 36 researchers seems adequate, especially in view of the rich qualitative data shared via the open-ended questions.

Nevertheless, we acknowledge that the mode of recruitment (convenience/snowball), the self-selection of eligibility to participate in the survey, and language/access limitations may have excluded some voices from the survey, for example, those of researchers working in non-Anglophone contexts. Additionally, we acknowledge that the experiences of the respondents may have been influenced by factors about which we did not enquire, such as the type of organisation for which they worked (e.g., humanitarian organisation, nongovernmental organisation or academic institute), each of which may have different types of support mechanisms in place; whether they worked as part of a team or sometimes on their own; and the types of local support and cooperation they had. While these factors may have influenced how the respondents experienced challenges, we believe it unlikely that they would have generated further challenges that were not identified in this study. Nevertheless, consideration of these factors could be useful for others who might contemplate the development of implementation guidance for the TRUST Code Supplement in future.

5.8 Conclusion

This survey study provided empirical validation and deeper elucidation of the wide-ranging challenges that researchers, participants and communities can face when conducting research in fragile settings. The findings demonstrate that such contexts amplify well-documented ethics and methodological issues while also generating unique risks, particularly to the wellbeing and safety of participants and researchers themselves. Importantly, the insights shared by respondents not only confirm the relevance of challenges identified in the literature but also enrich understanding of how they are experienced in practice and the strategies used to navigate them. While the study was not without limitations, the breadth and richness of the data contributed to a robust and credible evidence base that served as a foundation for the development of the TRUST Code Supplement.

References

Abimbola S (2019) The foreign gaze: authorship in academic global health. BMJ Glob Health 4:e002068. https://doi.org/10.1136/bmjgh-2019-002068

Antouly J, Leger V, Raillon C, Troit V (2024) The challenges of research in the humanitarian sector: an evolving relationship. Int Rev Red Cross 106(926):525–541. https://doi.org/10.1017/S1816383124000523

Blanchet K, Ramesh A, Frison S et al (2017) Evidence on public health interventions in humanitarian crises. Lancet 390(10109):2287–2296. https://doi.org/10.1016/S0140-6736(16)30768-1

Chatfield K, Schroeder D, Guantai A et al (2020) Preventing ethics dumping: the challenges for Kenyan research ethics committees. Res Ethics 17(1):23–44. https://doi.org/10.1177/1747016120925064

Chatfield K, Singh M (2025) Research ethics and integrity during pandemics: not unique, but vastly magnified challenges. In: Chatfield K, Singh M (eds) Research ethics and integrity during pandemics: developing the PREPARED code. SpringerBriefs in research and innovation governance. Springer, Cham, pp 1–7. https://doi.org/10.1007/978-3-031-91324-2

Diaconu K, Falconer J, Vidal N et al (2020) Understanding fragility: implications for global health research and practice. Health Policy Plan 35(2):235–243. https://doi.org/10.1093/heapol/czz142

Idris I (2019) Doing research in fragile contexts: literature review. GSDRC Research Report No. 1444. University of Birmingham, Birmingham. https://gsdrc.org/wp-content/uploads/2019/04/Doing-Research-in-Fragile-Contexts-Idris-2019.pdf Accessed 2 Sept 2025

Jansen YJ, Foets MM, de Bont AA (2010) The contribution of qualitative research to the development of tailor-made community-based interventions in primary care: a review. Eur J Pub Health 20(2):220–226. https://doi.org/10.1093/eurpub/ckp085

Kaur P, Stoltzfus J, Yellapu V (2018) Descriptive statistics. Int J Acad Med 4(1):60–63. https://doi.org/10.4103/IJAM.IJAM_7_18

Leresche E, Truppa C, Martin C et al (2020) Conducting operational research in humanitarian settings: is there a shared path for humanitarians, national public health authorities and academics? Confl Heal 14:25. https://doi.org/10.1186/s13031-020-00280-2

PREPARED (2025) The PREPARED Code—a global code of conduct for research during pandemics https://prepared-project.eu/prepared-code/. Accessed 10 Sept 2025

Rahi S, Alnaser FMI, Ghani MA (2019) Designing survey research: recommendation for questionnaire development, calculating sample size and selecting research paradigms. In: Economic and social development: book of proceedings, pp 1157–1169. https://www.proquest.com/conference-papers-proceedings/designing-survey-research-recommendation/docview/2188521137/se-2. Accessed 10 Sept 2025

Schroeder D, Chatfield K, Singh M et al (2019) Equitable research partnerships: a global code of conduct to counter ethics dumping. Springer Nature, Cham. https://doi.org/10.1007/978-3-030-15745-6

Schwandt TA (2014) The Sage dictionary of qualitative inquiry. Sage Publications, Thousand Oaks

Shanks K, Paulson J (2022) Ethical research landscapes in fragile and conflict-affected contexts: understanding the challenges. Res Ethics 18(3):169–192. https://doi.org/10.1177/17470161221094134

Tenzek K (2017) Snowball subject recruitment. In: Allen M (ed) The Sage encyclopedia of communication research methods, vol 4. SAGE Publications, Inc., Thousand Oaks. https://doi.org/10.4135/9781483381411.n569

Woodward A, Sheahan K, Martineau T, Sondorp E (2017) Health systems research in fragile and conflict affected states: a qualitative study of associated challenges. Health Res Policy Syst 15(1):44. https://doi.org/10.1186/s12961-017-0204-x

Wright S, O'Brien BC, Nimmon L et al (2016) Research design considerations. J Grad Med Educ 8(1):97–98. https://doi.org/10.4300/JGME-D-15-00566.1

Chapter 6
The TRUST CODE SUPPLEMENT—A Global Code of Conduct for Research in Fragile Settings

Abstract This chapter consists of the TRUST Code Supplement—A Global Code of Conduct for Research in Fragile Settings.

Keywords Research ethics · Research integrity · Research ethics in fragile settings · Research ethics in conflict-affected areas · Humanitarian research

Introduction

Research has the potential to help address humanity's challenges. This is especially vital in fragile settings such as disaster-affected areas, conflict or post-conflict zones, or high-crime informal settlements. In these contexts, where people often face threats, severe deprivation and limited access to basic services, their immediate focus is on survival, safety and wellbeing, rather than contributing to research.

While *The TRUST Code—A Global Code of Conduct for Equitable Partnerships in Research* (2018) supports efforts to overcome the systemic legacies of exclusion and unfairness in international research, fragile settings pose unique and complex challenges that require additional care and consideration. Accordingly, the TRUST Code forms the foundation for research in fragile settings, while this Supplement furthers its applicability in these contexts.

This *TRUST Code Supplement—A Global Code of Conduct for Research in Fragile Settings* (2025) is based on literature reviews, scoping reviews in English and German, and a survey study. The results were refined through extensive consultations with stakeholders.

The code:

- presumes adherence to the TRUST Code
- provides support across all research disciplines
- presents concise statements in clear language to encourage access
- combines guidance on research ethics and research integrity
- links each guidance article to the value of care, respect, fairness and honesty.

D. Schroeder et al., *Research Ethics and Integrity in Fragile Settings*, SpringerBriefs in Research and Innovation Governance,
https://doi.org/10.1007/978-3-032-36141-7_6

Care

Article 1

The **physical integrity, mental health and safety** of people in fragile settings should not be worsened through their involvement in research.

Article 2

Researchers must familiarise themselves with ongoing **rescue and relief operations** at the research site to ensure that their activities do not hamper or disrupt these efforts.

Article 3

Researchers must anticipate the risk of **retraumatising** research participants and local support teams (e.g., interpreters). Where retraumatisation is a possibility, trauma-informed practices (such as the appropriate training of researchers) should be implemented.

Article 4

Prior to the start of any field activities, researchers should be supported by their institution, wherever possible, to develop **risk management plans** to protect the physical integrity, mental wellbeing and safety of researchers, research participants and support teams. These plans should be tailored to the specific environment and, where feasible, include input from local collaborators.

Article 5

Research ethics committees should engage supportively with researchers in seeking ways to facilitate responsible research in fragile settings.

Article 6

It is the researchers' obligation to ensure that potential research participants understand that research activities are separate from humanitarian work (**'humanitarian misconception'**). Where members of the research team have roles in both areas, they must clearly explain the distinction to potential research participants.

Article 7

Conducting research in unstable environments can present ongoing challenges. Researchers must ensure that **adjustments to the research protocol**—in response to changing conditions—are locally appropriate and maintain the integrity of the research.

Article 8

The potential for **misuse of research *data*** can be high for research undertaken in fragile settings, especially in areas of conflict. Additional data security measures must be implemented including, wherever possible, methods that avoid the processing of personal data.

Article 9

The potential for **misuse of research *findings*** can be high for research undertaken in fragile settings, especially in areas of conflict. Dissemination should prioritise the needs and safety of local communities by, for instance, avoiding the release of information that might escalate local conflict.

Respect

Article 10

Where **local ethics review**, i.e., review within the host country, is not possible, justification for undertaking research without local ethics approval must be provided to the research ethics committee that approves the research.

Article 11

To avoid **cultural misunderstandings**, researchers working in fragile settings should collaborate with others who are familiar with the cultural context (e.g., senior researchers, mentors or community researchers) and liaise as closely as possible with local actors.

Article 12

Where standard **informed consent processes** are impractical or too risky, **reasonable adaptations** should be discussed and agreed with the relevant research ethics committee(s) and local collaborators.

Article 13

Local knowledge and community acceptance are essential to conducting research that is respectful and not patronising. Researchers should ensure that research participants and local communities have opportunities to contribute local knowledge and perspectives meaningfully.

Fairness

Article 14

Wherever possible, and in particular in internationally funded research, **local stakeholders** should be included in **research decision-making** to ensure that the focus and implementation of research are not driven solely by external research actors.

Article 15

Communication between research teams is essential to avoid the **unnecessary duplication of studies**, which can lead to a waste of resources and excessive burdens on some participants and communities and the underrepresentation of others.

Article 16

Research participants' time is precious when sustaining life and health takes priority. It is therefore important that the envisaged research promises to be of good enough quality to lead to meaningful results. If the **integrity of the research** is likely to be compromised, people must not be burdened with participation.

Article 17

In extremely fragile situations, payments or other benefits for participation in research can pose a significant risk of **undue inducement**. A balance must be achieved between avoiding undue inducement and preventing the exploitation of research participants and local members of the research team (e.g., gatekeepers, interpreters and drivers).

Article 18

Where local members of the research team lack adequate **protection mechanisms such as insurance**, researchers from high-income settings should urge their institution to extend the protections they enjoy to the whole team.

Article 19

Providing feedback to participants is good participatory practice but can be challenging in unstable environments. Researchers should plan for this in advance by, for instance, informing participants where they will be able to access results (e.g., via a website).

Honesty

Article 20

Where research would require **collusion** with non-research actors through **corruption**, the research must not take place.

Article 21

Researchers must be wary of raising **unrealistic expectations** among all local stakeholders. Managing expectations is a critical aspect of conducting research in fragile settings.

Article 22

Ahead of any research, researchers must clarify their options for handling **unexpected findings or human rights abuses** encountered during fieldwork. These options must be adapted to each local situation.

Article 23

Researchers who work in fragile settings may encounter **conflicts of interest** between, for instance, a research aim set by external funders (e.g., the aim of understanding the causes of migration from conflict areas, where the funder is a government that wants to reduce immigration) and the prime research needs of research participants. Researchers must ensure full transparency about potential conflicts of interest to avoid the exploitation of research participants.

The code was drafted as part of the PREPARED project under the lead author Prof. Doris Schroeder.

Chapter 7
Ethical Research in Fragile Settings: A Conclusion

Abstract Research in fragile settings poses significant challenges, yet the insights generated through research can improve future support efforts and help to make local systems more resilient. When conducted ethically, research can empower local communities, enhance outcomes and ensure that interventions are effective. By contrast, unethical research can intensify local tensions and hostilities, engender distrust and lead to more harm than benefit. The TRUST Code Supplement is a concise, accessible ethics code that is focused on fragile settings. It is based upon real-world risk analysis, refined through broad consultations, designed to be relevant across disciplines and authored by a global consortium. Like the TRUST Code, it is framed around the global moral values of fairness, respect, care and honesty. This concluding chapter presents a case vignette from research undertaken in climate-change-ravaged Kiribati to illustrate how the code, article by article, can help to make research more trustworthy and more equitable. Working with local communities and research participants in an equitable and humble manner fosters trust, which in turn encourages openness, honesty and collaboration, all crucial for excellence in research.

Keywords Research ethics · Research integrity · Fragile settings · Disaster · Conflict · TRUST Code

7.1 Introduction

Research in "fragile settings is challenging but offers transformative potential" (Bedingar 2025). Evidence-based insights can inform future activities, from rescue and support operations to policies that strengthen local systems to improve resilience. *Ethical* research can empower local communities, improve outcomes and guide interventions towards more effectiveness. *Unethical* research can worsen local hostilities, create distrust and lead to more harm than benefit.

D. Schroeder et al., *Research Ethics and Integrity in Fragile Settings*, SpringerBriefs in Research and Innovation Governance,
https://doi.org/10.1007/978-3-032-36141-7_7

To promote the former and avoid the latter, the TRUST Code Supplement provides guidance for researchers who work in fragile settings such as disaster-affected areas, conflict or post-conflict zones, and high-crime informal settlements.

Like other TRUST Family Ethics Codes (TRUST n.d.), the TRUST Code Supplement has seven distinctive features (Fig. 7.1). It is:

1. short and accessible, enabling all research stakeholders, including those who might be at risk of harm through research, to understand and apply it
2. targeted in scope and focus, namely fragile settings in this case
3. based upon real-world risks, having been drafted after risk analysis to ensure that the guidance is closely aligned with practical realities
4. broadly verified, having been refined through wide-ranging consultations
5. cross-disciplinary and suitable across research fields from archaeology to zoology
6. globally appropriate, having been authored by a global consortium
7. values-based, being grounded in the moral values of fairness, respect, care and honesty, which have been shown to be globally accepted but lacking in equitable research (Schroeder et al. 2019; Chatfield and Singh 2025).

This book has presented the research underpinning the TRUST Code Supplement, drawing on three main sources: a focused scoping review of English sources, a scoping review–informed evidence mapping of German sources and a survey of researchers who had first-hand experience of working in fragile contexts. Together, these components provided the empirical evidence base for the code.

Analysis of these findings revealed the key challenges to be addressed, the moral values at risk of being compromised, and the actions required to mitigate these challenges. The resulting ethics code was subsequently refined through extensive consultation with a broad range of stakeholders.

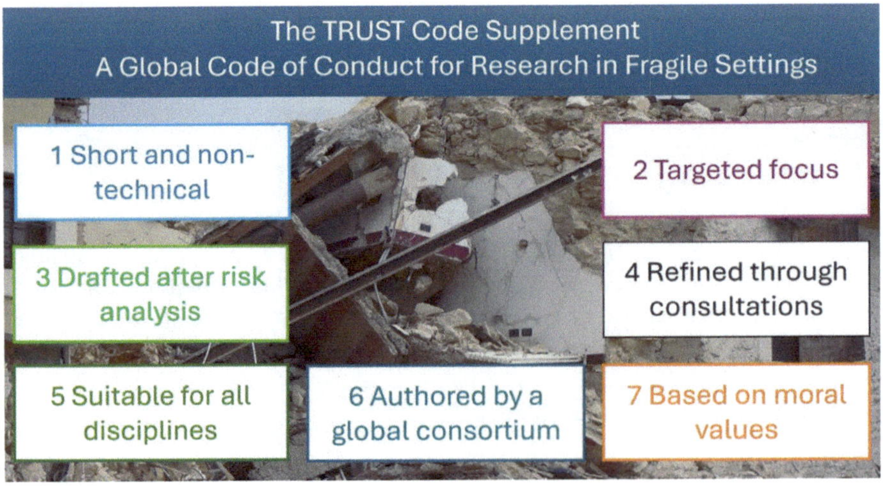

Fig. 7.1 Key features of the TRUST Code Supplement

Ethics guidance is key to the ethical conduct of research, both to assist researchers and research ethics committees, and also to convey what ethical research should look like to research participants and other stakeholders.

While ethics guidance and institutional support and oversight are important components for ensuring ethical research, the conduct and attitudes of individual researchers should never be underestimated (Frankel 1993; Shadare 2021). As our research and our own experience of conducting research in fragile settings have shown, ad hoc decisions have to be made much more often in fragile than in other settings, and being able to follow study protocols in minute detail can be the exception, not the rule.

We have seen in Chaps. 3, 4 and 5 what the multiple challenges for researchers in fragile settings are. In Chap. 2, we accompanied a researcher into Syria and to interactions with Syrian refugees in Germany. What came through was that difficulties can appear from the most unforeseen places: for instance, in obtaining research ethics approval.

The authors of this book have great respect for Young Su Park, one of our co-authors, who took risks and accepted very significant discomfort to obtain health data to improve people's lives in a setting ravaged by climate change. To illustrate how the TRUST Code Supplement might be applied in a real-world research project, the following case vignette draws upon his experience of conducting research in that particular setting.

7.2 A Kiribati Case of Responsible Research in a Fragile Setting

In 2024, Dr. Young Su Park undertook field research for a World Health Organization (WHO) climate change and health programme on the outer islands of Kiribati. The aim was to assess climate change risk perception and identify culturally acceptable climate and health resilience-building interventions.

The islands of Kiribati are located in the South Pacific near the equator between Hawaii and Australia. They consist of 32 atolls scattered over vast territorial seawaters equivalent in size to the contiguous mainland of the United States.

Kiribati is at extreme risk from climate change, facing an existential threat from rising sea levels that contaminate fresh water, destroy crops and threaten to submerge its low-lying islands, potentially making those islands uninhabitable. Extreme weather, drought and salinisation compound issues with food security, health challenges and displacement (WHO 2024).

This account highlights good practice examples from Young Su's research. Young Su is a medical doctor and anthropologist from South Korea who studies the ramifications of history and trauma in the field of global health. Currently he is Associate (Professor) in Medical Humanities at Seoul National University. He wrote this vignette, so it is in the first person. Another medical anthropologist, Prof. Taewoo

Kim, also from South Korea, was involved in the field research, but for ease of reading, this account is written as if by a single author ("I"), wherever the activity described heavily involves Young Su.

This vignette responds to the Trust Code Supplement by relating Young Su's work in Kiribati to each article in turn.

7.2.1 *The physical integrity, mental health and safety of people in fragile settings should not be worsened through their involvement in research.*

Planning the trip to Kiribati, the first question I asked myself was about the carbon footprint: "Am I worth the trip?" For the round trip to Marakei (one of the islands that was chosen for the research), eight flights would be necessary, covering a distance of 23,000 km from my university in Seoul. The trip was intended for the initial assessment of the larger *Te Mamauri* project for a climate-resilient health system in Kiribati. As the research aimed at improving the situation locally, and as no local researchers were available, I took the decision to travel.

Kiribati islanders suffer from the triple burden of communicable diseases, non-communicable diseases and their concomitant exacerbation by climate change (Kim and McIver 2015). Limited water supply and infrastructure for hygiene and sanitation have exposed them to heightened risks of waterborne diseases.

As the environmental footprint from the travel could not be reduced, I decided not to leave any recyclable waste in Kiribati and to take it all back with me on the return flights. Kiribati has limited space for landfills due to the narrowness of the atolls and is remote from any possible destination for exported waste (MELAD 2020). A landfill site in South Tarawa is at risk of sliding into the ocean due to storms and extreme weathers. The lack of waste management systems in Kiribati is connected to the burden of water- and vector-borne diseases, one of the reasons for the research, and taking all recyclable waste home was one way of making sure that the research had no negative side effects.

7.2.2 *Researchers must familiarise themselves with ongoing rescue and relief operations at the research site to ensure that their activities do not hamper or disrupt these efforts.*

Prior to the research, I reached out to NGO workers (Médecins Sans Frontières, Red Cross) in Kiribati, to understand existing operations and programmes in Kiribati. These connections also allowed me to share relevant findings from the field studies later.

7.2.3 *Researchers must anticipate the risk of retraumatising research participants and local support teams (e.g., interpreters). Where retraumatisation is a possibility, trauma-informed practices (such as the appropriate training of researchers) should be implemented.*

When I was working at Haverford College in the USA, I learned about the importance of addressing and preventing secondary trauma at a trauma-informed approach workshop. There I also designed and taught a course titled "Trauma, Historical Memory, and Embodiment".

In Kiribati, I was indeed concerned about the possibility of retraumatisation because the current climate crisis is severe, also leading to significant migration. On my first trip, I learned that I-Kiribati islanders had suffered from environmental destruction catalysed by colonial resource extractions and nuclear experiments conducted by US and UK forces. I therefore, in order to be more aware of the exploitative background, studied the history of Kiribati in more detail before returning for subsequent field visits. This, I hope, allowed me to be more sensitive in my field studies.

7.2.4 *Prior to the start of any field activities, researchers should be supported by their institution, wherever possible, to develop risk management plans to protect the physical integrity, mental wellbeing and safety of researchers, research participants and support teams. These plans should be tailored to the specific environment and, where feasible, include input from local collaborators.*

Before the field study commenced, I had a meeting with governance staff at my university to discuss risk management plans for health and safety.

7.2.5 *Research ethics committees should engage supportively with researchers in seeking ways to facilitate responsible research in fragile settings.*

This field study received ethics approval from the institutional review board of Seoul National University Hospital. One of the reviewers raised reasonable questions about the local contact and translations into local languages. In general, the committee was supportive.

7.2.6 *It is the researchers' obligation to ensure that potential research participants understand that research activities are separate from humanitarian work ("humanitarian misconception"). Where members of the research team have roles in both areas, they must clearly explain the distinction to potential research participants.*

As there was no immediate disaster situation, there were no activities that could be regarded as humanitarian rescue operations.

7.2.7 *Conducting research in unstable environments can present ongoing challenges. Researchers must ensure that adjustments to the research protocol—in response to changing conditions—are locally appropriate and maintain the integrity of the research.*

While I was in Kiribati, it became clear that follow-up studies would be necessary. A colleague and I presented these as new studies, rather than adjustments to the research protocol. We experienced significant bureaucratic delays in preparing follow-up studies and subsequent projects.

7.2.8 *The potential for misuse of research data can be high for research undertaken in fragile settings, especially in areas of conflict. Additional data security measures must be implemented including, wherever possible, methods that avoid the processing of personal data.*

I collected basic demographic data during household visits. Personal data were anonymised and de-identified between my taking field notes and putting them into my laptop, and subsequent research outputs in the form of journal articles, reports and academic presentations. Since we did not anticipate misuse of research data, we deemed these standard data protection measures adequate.

7.2.9 *The potential for misuse of research findings can be high for research undertaken in fragile settings, especially in areas of conflict. Dissemination should prioritise the needs and safety of local communities by, for instance, avoiding the release of information that might escalate local conflict.*

There is a risk of misrepresentation of what is happening in Kiribati, in particular by mainstream media in the Global North. The risk of rising sea levels from climate change is being overemphasised in the representation of Kiribati, since atolls are only 2 or 3 m above sea level anyway. I was much more concerned about the underestimation of other, more devastating environmental harm to I-Kiribati islanders: in particular, droughts and extreme weather events catalysed by global climate change. I made every effort in my dissemination to deliver a realistic picture of the climate impact as witnessed and experienced by I-Kiribati islanders.

7.2.10 *Where local ethics review, i.e., review within the host country, is not possible, justification for undertaking research without local ethics approval must be provided to the research ethics committee that approves the research.*

I could not secure local ethics review approval due to the absence of research ethics review committees in Kiribati. Instead, I received a research permit from the government of Kiribati after they reviewed the research proposal.

7.2.11 *To avoid cultural misunderstandings, researchers working in fragile settings should collaborate with others who are familiar with the cultural context (e.g., senior researchers, mentors or community researchers) and liaise as closely as possible with local actors.*

I benefited a lot from cultural insider perspectives of I-Kiribati translators: a schoolteacher, a government officer and a local staffmember of the WHO. For instance, following their advice, I paid respect to four spirits of Marakei island through a visitors' ritual tour around the island upon my arrival.

7.2.12 *Where standard informed consent processes are impractical or too risky, reasonable adaptations should be discussed and agreed with the relevant research ethics committee(s) and local collaborators.*

It was reasonable to obtain written informed consent with signatures, and therefore no adaptation to this standard method was necessary.

7.2.13 *Local knowledge and community acceptance are essential to conducting research that is respectful and not patronising. Researchers should ensure that research participants and local communities have opportunities to contribute local knowledge and perspectives meaningfully.*

My participation in the communal dancing of the *Karinang* (welcome ceremony) at a community gathering (*Maneaba,* meaning a big house) laid the foundation for community acceptance among Marakei islanders. I also showed respect by learning about herbal medicine and massage from traditional healers, in addition to gathering the information I had come for (on climate change impact). Most importantly, though, I tried, throughout the fieldwork, to capture authentic views from the I-Kiribati islanders on the impact of climate change without exposing them to preconceived ideas. What *they* had to tell me was of prime importance.

7.2.14 *Wherever possible, and in particular in internationally funded research, local stakeholders should be included in research decision-making to ensure that the focus and implementation of research are not driven solely by external research actors.*

The main focus of the research was driven by the WHO. However, I had a kick-off meeting with Marakei island council members to explain the purpose of the field study and to seek guidance and suggestions from local stakeholders.

7.2.15 *Communication between research teams is essential to avoid the unnecessary duplication of studies, which can lead to a waste of resources and excessive burdens on some participants and communities and the underrepresentation of others.*

I conducted a thorough literature review to avoid the unnecessary duplication of studies. There were very few studies on the health and culture of Kiribati.

7.2.16 *Research participants' time is precious when sustaining life and health takes priority. It is therefore important that the envisaged research promises to be of good enough quality to lead to meaningful results. If the integrity of the research is likely to be compromised, people must not be burdened with participation.*

To check the quality of our research as frequently as possible, and while the research was going on, I shared my field notes with another anthropologist and had daily review meetings to check the progress of interviews and observations.

7.2.17 *In extremely fragile situations, payments or other benefits for participation in research can pose a significant risk of undue inducement. A balance must be achieved between avoiding undue inducement and preventing the exploitation of research participants and local members of the research team (e.g., gatekeepers, interpreters and drivers).*

With my colleague, I conducted 53 in-depth household interviews and interviews with healthcare providers. An important ethical question was how to compensate the people who agreed to take part in the interviews.

I consulted with local I-Kiribati research collaborators about reasonable and respectful compensation for interviewees, and also with university research administrators in Korea. In South Korea, qualitative research participants receive around 100 Australian dollars (AUD) (100,000 Korean won) for a 1-h interview. The government of Kiribati provides outer islanders in Kiribati with roughly 200 AUD per month as compensation for their copra (dried coconut husk) production. In 2024 South Korea's gross domestic product per capita was $61,051 in international dollars (World Bank n.d.b). The equivalent for Kiribati was $3,702 in international dollars (World Bank n.d.a). After the matter had been carefully considered, each household received 20 AUD (about 12 euros) for an hour-long interview out of the 50,000 USD (about 42,000 euros) budgeted for the entire year-long research project.

However, a European colleague working for an international organisation criticised our compensation for I-Kiribati research participants. According to her, we were not supposed to provide any monetary remuneration because that set an unnecessary precedent that would make local residents expect financial compensation for any future research in Kiribati.

We believed there should be reasonable compensation for the time, knowledge and effort research participants shared with researchers.

7.2.18 *Where local members of the research team lack adequate protection mechanisms such as insurance, researchers from high-income settings should urge their institution to extend the protections they enjoy to the whole team.*

This article is particularly important for fieldwork that involves risks for local teams. For instance, in 2024, a UK-Nepal collaboration involved in post-disaster heritage protection following Nepal's 2015 earthquake checked whether their collaboration was equitable in terms of he articles of the TRUST Code (Coningham et al. 2024). They found that insurance cover for the Nepal-based collaborators, who were not insured through their organisation, would have made the research even more equitable (ibid). This article was not applicable to the Kiribati research, however, because there were no identifiable risks for interviewees that would have required insurance cover.

7.2.19 *Providing feedback to participants is good participatory practice but can be challenging in unstable environments. Researchers should plan for this in advance by, for instance, informing participants where they will be able to access results (e.g., via a website).*

I had a wrap-up meeting with the Marakei island council to share the findings from our study and to listen to the interpretation of research findings by local stakeholders. They also shared the findings with nurses working in the health centre who participated in interviews.

7.2.20 *Where research would require collusion with non-research actors through corruption, the research must not take place.*

At no point in the research did collusion with local stakeholders or bribes for local actors become a topic.

7.2.21 *Researchers must be wary of raising unrealistic expectations among all local stakeholders. Managing expectations is a critical aspect of conducting research in fragile settings.*

I informed research participants that their recommendations would not necessarily be implemented. I explained that their suggestions would be reviewed and deliberations by the WHO and the government of Kiribati would then follow.

7.2.22 *Ahead of any research, researchers must clarify their options for handling unexpected findings or human rights abuses encountered during fieldwork. These options must be adapted to each local situation.*

I was concerned about the possibility of encountering victims of domestic violence or sexual violence based on previous research conducted in Kiribati. I identified potential sources of help in advance (e.g., the Kiribati Family Health Association and local healthcare providers), but I did not observe any cases during the field study.

7.2.23 *Researchers who work in fragile settings may encounter conflicts of interest between, for instance, a research aim set by external funders (e.g., the aim of understanding the causes of migration from conflict areas, where the funder is a government that wants to reduce immigration) and the prime research needs of research participants. Researchers must ensure full transparency about potential conflicts of interest to avoid the exploitation of research participants.*

The aim of the research, as set by the research funder (the WHO), was fully to the benefit of Kiribati. There was no conflict of interest.

While this case study showed how the TRUST Code Supplement can be employed by researchers in one particular fragile setting, Table 7.1 summarises possible uses of the Supplement by different stakeholders.

Table 7.1 Possible use of TRUST Code Supplement by stakeholders

Stakeholder	Possible use of supplement
Research teams and consortia	Guiding ethical protocol development, fieldwork planning and all stages of project implementation Clarifying roles, authorship and equitable collaboration practices
Research ethics committees	Informing ethics review processes and frameworks. Reflecting prompts during reviewing activities
Funders	Formulating funder expectations of researchers regarding due diligence, safeguarding and equitable collaboration practices
Community partners	Guiding expectations in terms of engagement, dissemination, benefits and protection against harms including misuse of research data and findings

7.3 Conclusion

By making his research locally relevant, tailoring it to local needs and focusing on listening, Young Su ensured that his interviewees would share many stories with him.

One story from an elderly man who invited Young Su to dinner remains particularly powerful. The elderly man asked him what he thought the starting point of all the ongoing tragedies was. Young Su thought that he was talking about the Anthropocene: that is, the time when humans started to have a substantial impact on the planet. So he began to recall the Industrial Revolution of the eighteenth century. However, the elderly man disagreed and said that it all began after World War II. He said: "Around that time, UK and USA [forces] bombed Christmas Island for nuclear experiments. And it was not uninhabited islands. People lived there, when it was bombed. They think that it's okay to bomb us because we were mere savages for them."

Young Su was taken aback by this answer and looked up the history after dinner. Indeed, 33 nuclear experiments by UK and USA forces occurred in 1957 and 1962 (Hill 2024). A communication from London during that time responded to a concern about people living there. It said: "There are savages" (Walsh 2020). The elderly man said that through climate change we were throwing a "nuclear bomb" at the I-Kiribati people every day, as if there was no one living there, as if they were not the same humans as we were. Every day we were throwing a "nuclear bomb" in the form of carbon emissions onto I-Kiribati people, who were seemingly invisible to us. The elderly man suggested that we should instead think about what we should do, rather than what *they* were supposed to do to adapt to risks from the climate crisis.

The global teams involved in the TRUST Family of Ethics Codes (SASI 2017; TRUST 2018; PREPARED 2025a; PREPARED 2025b), believe strongly that research which satisfies the values of fairness, respect, care and honesty will be better research—not only more authentic, but also characterised by higher excellence and a higher potential for impact. That is because working equitably and with humility with local communities and local research participants creates trust, which in turn creates openness, honesty and collaboration, which are crucial for excellent research. We hope that the case vignette from Dr. Young Su Park's research shows that it is possible, reasonable and beneficial for all involved to adhere to the TRUST Code Supplement.

References

Bedingar E (2025) Breaking the aid trap: why fragile states need local solutions, not global fixes. PLOS Glob Public Health, 2 January. https://speakingofmedicine.plos.org/2025/01/02/breaking-the-aid-trap-why-fragile-states-need-local-solutions-not-global-fixes/. Accessed 25 Feb 2026

Chatfield K, Singh M (2025) Research ethics and integrity during pandemics: developing the PREPARED Code, SpringerBriefs in Research and Innovation Governance. Springer, Cham. https://doi.org/10.1007/978-3-031-91324-2

Coningham R, Lewer N, Acharya KP et al (2024) Enabling equitable and ethical research partnerships in crisis situations: lessons learned from post-disaster heritage protection interventions following Nepal's 2015 earthquake. Res Ethics 20(4):835–846. https://doi.org/10.1177/17470161241234502

Frankel MS (1993) Professional societies and responsible research conduct. In responsible science: ensuring the integrity of the research process, vol 2. National Academy Press, Washington, DC, pp 26–49. https://www.ncbi.nlm.nih.gov/books/NBK236199/. Accessed 25 Feb 2026

Hill C (2024) 'Our nuclear childhood': the sisters who witnessed H-bomb tests on their Pacific island and are still coming to terms with the fallout. The Conversation, 24 October. Accessed 25 Feb 2026

Kim R, McIver L (2015) Human health and climate change in Pacific Island countries. World Health Organ. https://www.who.int/publications/i/item/human-health-and-climate-change-in-pacific-island-countries. Accessed 25 Feb 2026

MELAD (2020) Kiribati waste management resource recovery strategy 2020–2030. Ministry of Environment, Lands and Agricultural Development, Kiribati. https://pacific-data.sprep.org/dataset/kiribati-waste-management-resource-recovery-strategy-2020-2030. Accessed 25 Feb 2026

PREPARED (2025a) The PREPARED Code—a global code of conduct for research during pandemics https://prepared-project.eu/prepared-code/. Accessed 10 Oct 2025

PREPARED (2025b) The TRUST Code Supplement—a global code of conduct for research in fragile settings. https://www.globalcodeofconduct.org/trust-code-supplement/. Accessed 10 Oct 2025

SASI (2017) San code of research ethics. South African San Institute, Kimberley. https://www.globalcodeofconduct.org/wp-content/uploads/2025/08/San-Code-of-RESEARCH-Ethics-Booklet_English.pdf. Accessed 25 Feb 2026

Schroeder D, Chatfield K, Singh M et al (2019) Equitable research partnerships: a global code of conduct to counter ethics dumping. Springer Nature, Cham https://doi.org/10.1007/978-3-030-15745-6

Shadare GA (2021) Managing ethical tensions when conducting research in fragile and conflict-affected contexts. In: Kara H, Khoo S (eds) Qualitative and digital re-search in times of crisis: methods, reflexivity and ethics. Bristol University Press, Bristol, pp 218–234

TRUST (2018) The TRUST Code: a global code of conduct for equitable research partnerships. https://doi.org/10.48508/GCC/2018.05

TRUST (n.d.) The TRUST family of ethics codes. https://www.globalcodeofconduct.org/. Accessed 25 Feb 2026

Walsh MR (2020) A history of Kiribati: from the earliest times to the 40th anniversary of the republic. Independently published

WHO (2024) Countering the climate crisis in Kiribati. 8 March. World Health Organization. https://www.who.int/westernpacific/newsroom/feature-stories/item/countering-the-climate-crisis-in-kiribati. Accessed 25 Feb 2026

World Bank (n.d.a) GDP per capita, PPP (current international $)—Kiribati. https://data.worldbank.org/indicator/NY.GDP.PCAP.PP.CD?locations=KI. Accessed 25 Feb 2026

World Bank (n.d.b) GDP per capita, PPP (current international $)—Korea, Rep. https://data.worldbank.org/indicator/NY.GDP.PCAP.PP.CD?locations=KR. Accessed 25 Feb 2026

Glossary

Displaced persons Displaced persons are individuals compelled to leave their homes due to conflict, human rights violations or environmental disasters.

Ethics dumping Ethics dumping is the practice of exporting unethical research from higher- to lower-income regions, where ethics oversight and governance mechanisms may be less well-resourced or regulated.

Forced migration Forced migration is the non-voluntary relocation of individuals or communities driven by conflict, human rights violations or environmental disasters.

Fragile settings Fragile settings are environments where significant risks intersect with an insufficient capacity to manage, mitigate or recover from those risks. For the TRUST Code Supplement, we focused upon disaster-affected areas, conflict or post-conflict zones, and high-crime informal settlements.

Helicopter research Helicopter research refers to exploitative studies where researchers from higher-income settings extract data or other resources from lower-income settings without involving local researchers or sharing research benefits fairly. It is often characterised by fleeting visits from foreign researchers and limited, superficial engagement with local communities and research participants.

Humanitarian misconception A humanitarian misconception is the mistaken belief that participation in research is necessary to receive essential humanitarian aid.

Protection-Inclusion Dilemma The protection-inclusion dilemma is a major challenge in research ethics. It describes the tension between protecting people in vulnerable situations from potential harm through exclusion and promoting their inclusion so they can contribute to and benefit from the interventions and innovations generated by research, despite the risks that participation may entail.

Retraumatisation Retraumatisation can occur when research participants are asked to discuss or recount distressing experiences, causing them to relive aspects of their past trauma.

D. Schroeder et al., *Research Ethics and Integrity in Fragile Settings*,
SpringerBriefs in Research and Innovation Governance,
https://doi.org/10.1007/978-3-032-36141-7

TRUST Family of Ethics Codes The TRUST Family of Ethics Codes provides an ethics framework that is designed to promote equitable research relationships in international research collaborations, in fragile settings, during pandemics and when working with the San community in South Africa.

Undue inducement Undue inducement refers to an incentive or reward that has the potential to undermine voluntary informed consent, when individuals are considering whether to participate in a research study.

FSC
www.fsc.org
MIX
Papier | Fördert
gute Waldnutzung
FSC® C083411

Kolibri 360 GmbH
Ferdinand-Jühlke-Straße 7
99095 Erfurt, Deutschland
produktsicherheit@kolibri360.de